D0278564

Inspection Machines, Measuring Systems & Instruments

Inspection Machines, Measuring Systems & Instruments

**H C TOWN &
H MOORE**

B.T.BATSFORD LTD London

© H C Town and H Moore 1978
First published 1978

ISBN 0 7134 0795 6 (cased edition)
     0 7134 0796 4 (limp edition)

Printed by Billing and Son Ltd
of London Guildford and Worcester
for the Publishers B T Batsford Ltd,
4 Fitzhardinge Street, London W1H 0AH.

CONTENTS

I.
Historical Aspects of Inspection & Measurement

Developments of Maudsley and Whitworth

From the earliest days, man in his struggle for material progress has been compelled to make measurements. In particular he has had to measure and standardize the fundamental units of time, mass and distance. Our concern in this book, however, is primarily with the measurement of length.

The great difficulties which James Watt encountered in obtaining accurately manufactured parts for his early steam engines are well known. Later similar problems had to be faced in the production of machinery for the textile and other industries. Each part was a separate piece of engineering, for there was no standardization or interchangeability. A visitor to Whitworth's works from *The Manchester City News* in 1865 pointed out that, 'when Mr Whitworth started business in Manchester in 1833, the best mechanics were accustomed, in common parlance, to speak of a "full eighth" of an inch, or a "bare sixteenth" ', the latter term expressing to their minds something like perfection in mechanical finish. Mr Whitworth's foot rule, on which he had thirty-secondths of an inch marked, was regarded as a curiousity, and many did not hesitate to affirm that to work to such a standard was an unnecessary refinement. How very different the case is now may be understood when we state that, at the works of Whitworth and Co. in Chorlton Street, the self-acting machines are made, adjusted and fitted to a ten-thousandth of an inch.'

Whitworth perfected many ideas introduced by Maudslay, for it was the latter who first recognized the importance of true plane surfaces, who produced surface plates by making three at a time and who checked them against each other for being either concave or convex. His method was by scraping instead of by filing until accuracy was achieved, but what was more important he had these standard planes put on the benches beside his workmen and insisted that they be used constantly to test their work. Maudslay also specialized in producing accurate screws and finally produced a splendid brass screw of 7 ft long which was only one-sixteenth of an inch in error of its computed length. This was an error of about 0.002 in. per thread, and he then produced a very clever linkage by means of which these small errors could be corrected.

The Measuring Machine

The measuring machine was one of Whitworth's major concerns, but his debt to Maudslay is again evident. Nasmyth states that Maudslay constructed a micrometer which he named the Lord Chancellor, in view of the absolute truth of its verdicts. At the time there was no standard measurement below the official yard, which was of little use in engineering. Thus Whitworth developed from Maudslay's micrometer his measuring machine which could measure to 0.0001 in. and a few years later made another capable of detecting differences of less than one-millionth of an inch. For this Whitworth received a medal at the Great Exhibition of 1851.

The practical use of the machine was in the production of standard gauges although Richard Roberts, another famous Manchester engineer, was using them as early as 1825 in his own works. Since Roberts had worked at Maudslay's before Whitworth, the system may have originated there. However, it was Whitworth in 1841 who advocated before the Institution of Civil Engineers 'the general use of standard gauges, graduated to a fixed scale, as a constant measures of size'. He pointed out that these would enable standard machine parts to be produced, instead of the existing 'indefinite multiplication of sizes'.

These developments led him to be a strong advocate of the decimal system, as it was impossible to work to small differences when using the fractional system and when measuring by sight from the lines of a rule. He demonstrated the accuracy obtainable by using his standard gauges, 'I have brought an internal gauge with a bore of 0.5770 in. diameter and two external cylinders, one being 0.5769 and the other 0.5770 in. diameter. The latter fits tightly in the internal gauge when both are clean and dry, while the smaller gauge is so loose as to appear not to fit at all.'

In a letter to *The Times* in 1855, Whitworth stated that the engines of ninety gunboats for the Crimean War were completed in 90 days, only as a result of the adoption of standard gauges so that intermatching parts could be produced in quantity by different firms. These arguments proved convincing, and in 1880 his standards were officially adopted by the Board of Trade. He is, of course, best known for his standardization of screw threads which was universally adopted by 1858, again after much pioneering work by Maudslay.

**Orders of
Accuracy**

The vast improvements in accuracy of engineering methods within the last 50 years tend to obscure the fact that optical instruments have for a much longer period been made to standards of accuracy higher than any yet achieved in normal engineering. It is only in comparatively recent years, however, that optical instruments have been at all widely applied in mechanical engineering. This does not mean that no developments were taking place, for certain progressive firms producing machines of high accuracy were using optical means. The production of an accurately divided circle was first accomplished by Marc Thury of the Swiss firm Société Genevoise of Geneva as long ago as 1882. The fact that any complete circle must total 360° enables sub-divisions of a circle to be made by a process of bisection known as the 'caliper principle'.

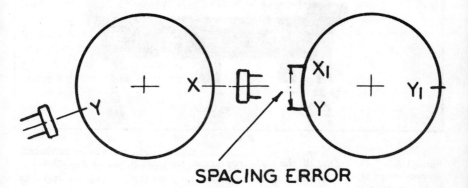

SPACING ERROR

Figure 1.1.
*Diagram showing
Thury's method of
dividing a circle*

Thury's method can be understood by reference to Figure 1.1. Two trial lines are marked at approximately 180°, and two measuring microscopes are then focused on the two lines X and Y. The circle is then rotated until the point Y occupies the position previously occupied by point X. Should the trial lines be situated at exactly 180° spacing, the microscope reading at the other side will be unchanged. Any error in the original 180° setting can be measured by means of the microscope at the left-hand side. If one of the trial lines is then moved a distance equal to half this reading, true 180° spacing will result. This process was continued until the master circle was divided into thirty-two equal parts with a precision of 1' of arc.

**Quality
Assurance**

As a result of the improvements made in instrumentation the inspector now has available a range of supporting equipment in the form of standards, measuring instruments and machines.

Figure 1.2.
*An early design of
inspection machine
(By courtesy of
DEA, Turin)*

One of the early inspection machines is shown in Figure 1.2. This was the first experimental prototype by Digital Electronic Automation S.P.A. (DEA) of Turin, Italy, and the present equivalents are described later. The machine was built in 1963 and was used as a basis for the design of later three-dimensional machines. The supporting equipment for all the present measuring machines is shown adjacent to the Olivetti machine, Figure 1.3.

The micro-computer can be used to obtain (1) the automatic alignment of the workpiece on the table by measuring the error signals, (2) the evaluation of the tolerances along the axes by calculating the differences between the measured values and the theoretical values (these are printed out and can be read from a tape reader), (3) the evaluation of the tolerance in the three places with a known diameter probe, the calculation of the diameter of the hole and the co-ordination of its centre, (4) the conversion of polar to cartesian co-ordinates or vice versa, (5) the calculation of different co-ordinates in relation to different zero points and (6) the punched tape generation for numerical control (NC) machining.

An increasing complexity of modern equipment is creating a demand for these machines, for higher standards of quality has led to a reduction in the tolerances to which parts must be made and consequently to a higher standard of inspection as part of the production process. The use of these modern measuring techniques has led, in one case, to the elimination of the expensive operation of grading such objects as gudgeon pins and pistons, and in another case to dispensing with the selective assembly of bevel gears and pinions.

Developments are proceeding with fibre optics to avoid heat radiation from light sources in measuring instruments. Complete in-process inspection of parts as they are machined is being investigated using sensors and instruments which are able to measure and control, continuously and almost instantaneously, the exact dimensional position and characteristics of the surfaces that are being generated by a cutting tool or grinding wheel. In later chapters, examples of such features on machines are given. Another important factor in accomplishing the highest accuracy in measurement is by feedback control to correct thermal deformation of machine structures. In addition, the laser interferometer for positioning the tables or saddles and workpieces on machine tools is available to give an accuracy of position of 1μm.

Figure 1.3.
The supporting equipment for a measuring machine:
A *peripheral control unit*
B *micro-computer*
C *memory system*
D *tape punch*
E *tape reader*
F *printing unit*
G *teletype*
H *tape punch*
(By courtesy of Olivetti, Italy)

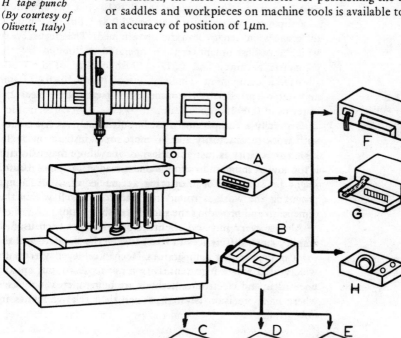

2.
Electrical, Electronic & Fluidic Measuring Systems

Comparison of Amplifying Systems

Before describing actual systems it is proposed to discuss the considerations that should be given to instrument features. Perhaps the most important is the magnification or amplitude of the instrument. It may be thought that the higher the sensitivity the better, but this is not the case. A high magnification usually restricts the range of the instrument; also a highly sensitive instrument reacts to any variation in ambient conditions.

A dial indicator may have a magnification of from 50:1 to 500:1, an inspection comparator 1,000:1, while the amplitude of a slip gauge comparator may be 50,000:1 but only minute differences in length can be determined. The accuracy to which any instrument can be operated is limited by the geometric accuracy and surface finish quality of the work. Clearly, a component which is out of round by 0.025 mm and out of parallel by the same amount cannot be measured in terms of 0.0025 mm.

Generally a comparator must be sufficiently precise to deal with work satisfactorily and no more so; an instrument of too high a sensitivity is just as useless as one whose magnification is too low. A rough indication is that a magnification of 1000:1 might be suitable for measuring accuracies of ±0.0025 mm providing the work is round and parallel to well within this dimension and providing the surface finish is good.

Almost every physical scientific principle has been used to amplify small differences in size between the standard and the work in engineering comparators. Mechanical lever systems are widely used where high sensitivity is not required, but optical, pneumatic and electronic methods are being increasingly used where high precision is essential; and their relative merits are now given.

MECHANICAL
Advantages
1 This is usually cheaper than other systems and independent

of air or electricity.

2 It is good for workshop conditions with a clear scale that is easily understood.

3 The system is usually compact and easily portable and can be issued from a store.

Disadvantages

1 There are many moving parts, tending to increase friction and to reduce accuracy.

2 The range is limited in any instrument with a pointer moving over a scale.

3 The mechanism has inertia which may make the instrument sensitive to vibration.

4 Errors of parallax can occur when the pointers move over the scales.

OPTICAL

Advantages

1 The small number of moving parts contributes to high accuracy.

2 The scale can move past a fixed datum line to give high magnification and no parallax errors.

3 If 2 is complied with, the operator observes the same part of the instrument dial irrespective of the size of the part. Also the optical lever is weightless.

Disadvantages

1 An electrical supply is necessary.

2 The instruments through which the scale is viewed through the eyepiece of a microscope are not convenient for continuous industrial use.

3 The apparatus is comparatively large and expensive.

4 On instruments with high magnification, the heat from the lamp and the transformer may cause the setting to drift.

PNEUMATIC

Advantages

1 There are a small number of moving parts and none in some cases.

2 This system has a very small measuring pressure, often an air jet only.

3 Very high magnification is possible, and the indicator can be remote from measuring unit.

4 This is very useful for measuring diameters when these are holes that are small compared with the length.

Disadvantages

1 Elaborate auxiliary equipment is necessary, including accurate pressure regulation.

2 The scale may be of liquid in a glass tube, and meniscus

errors may make high magnification necessary for high accuracy.

3 The apparatus is not easily portable and is elaborate for industrial uses.

ELECTRICAL
Advantages

1 The measuring unit can be remote from the indicating instrument.

2 This system has a high magnification with small number of moving parts.

3 The mechanism carrying the pointer is light and not sensitive to vibration.

4 On an a.c. supply, the cyclic vibration reduces errors due to sliding friction.

5 The measuring unit can be small, and the instrument can have several magnifications.

Disadvantages

1 Heating of coils in the measuring unit may cause zero drift and may alter the calibration.

2 A fixed scale with a moving pointer gives small range at high magnifications.

Electrical Measuring and Inspection Means

WHEATSTONE BRIDGE
Reference is again made to the Wheatstone bridge arrangement, for many electronic comparator instruments are based on a modification of this. The bridge is used by electrical engineers for the measurement of electrical resistance as indicated in Figure 2.1. If the bridge is to balance electrically, the ratio of the resistances in each pair must be equal, or

$$\frac{R_1}{R_2} = \frac{R_3}{R_4}$$

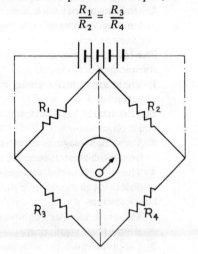

Figure 2.1.
A diagram showing the Wheatstone bridge arrangement

If this is the case, then there will be no current flowing through the galvanometer in the centre. If three of these resistances are of known value and if one can be varied, then the variable resistance can be adjusted until the bridge is balanced (shown by a zero galvanometer reading), and it is a simple matter to evaluate the unknown resistance. It is important to note that a Wheatstone bridge circuit using resistive elements can only be used with a d.c. supply.

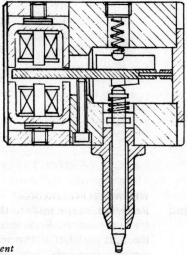

Figure 2.2.
An electrolimit gauging instrument

ELECTROLIMIT GAUGE

In this instrument an a.c. bridge is employed, and when alternating current is used the inductance and capacitance of the arms must be accounted for, in addition to their resistances. The lay-out of the head is shown in Figure 2.2. Movement of the plunger actuates an iron armature supported by a flexible-spring strip which causes a variation in the air gap between the armature and a pair of coils forming one pair of the arms of the a.c. bridge. The bridge is balanced when the air gaps are equal. If movement of the plunger causes the air gaps to differ, then the reluctance of the magnetic circuit at each side will be changed. This alters the inductance of the coils which is a measure of the magnetic effect produced when a current is passed through them. The balance of the bridge is thus upset, and a current will flow which can be accurately measured by means of a microammeter. The instrument is calibrated to read in linear units corresponding to the small displacements of the plunger when components are fed under the plunger.

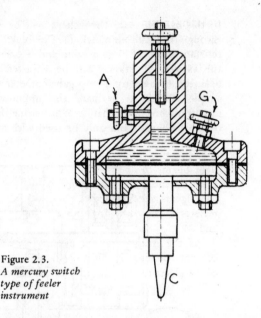

Figure 2.3.
*A mercury switch
type of feeler
instrument*

MERCURY SWITCH FEELER

There are many applications where the height of a column of mercury is used to make contact and thereby to give size indication or to operate the automatic sizing system of a grinding machine. Very sensitive control can be obtained by the type of feeler shown in Figure 2.3. The device contains mercury in a small bore at the top which broadens at the base so that if we assume that the bores are 6 and 60 mm respectively, and if this acts on the squares of the ratios of the diameter, this gives a sensitivity of 100:1 from the mercury alone; in fact the system can be regarded as a hydraulic travel amplifier. The screw A acts as a circuit breaker as the feeler C moves upwards and downwards. Apart from checking dimensions the device finds applications in die sinking; for this the feeler is fastened to the head of the milling machine by a rod and connecting arm, while current is supplied through the mercury by connection to the screw G so that, when a circuit is made through A, the vertical feed motor is reversed until the level of the mercury falls and the circuit is broken.

LEITZ EQUIPMENT

This equipment is made by E. Leitz, Wetzlar. A diagram showing the arrangement of a linear transducer is given in Figure 2.4. Light from the lamp A passes through the condenser lens B and the red filter C. The image of the scale is deflected by the pentaprism E through the Wollaston beam-splitting prism F, which produces two sets of scale images, displaced relative to each other and polarized at 90°. Both images are re-formed in an identical position on the scale surface by the action of the prism G and objective lenses H but are displaced relative to each other by 0.25 of the scale division width.

Because the optical path of the light forming both images is the same, measuring errors cannot be introduced. When the scale is moved longitudinally, the images of the scale travel in the direction opposite to that of the scale movement. The light beams composing the scale images also pass through the scale to the second beam-splitting prism K, which separates the signals according to their direction of polarization and reflects them through condenser lenses to the photo-cells L and M.

As the images traverse the scale markings, the intensity of light reaching the photo-cells is modulated, and the phases of the electrical signals produced cover 0.25 of a scale division

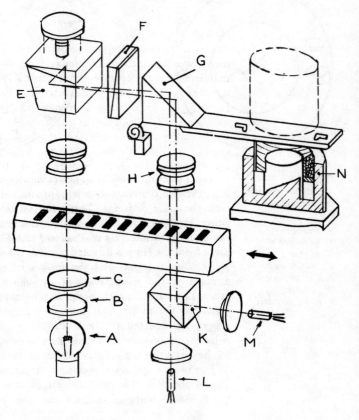

Figure 2.4.
*An optical system
using a linear
transducer:*

A *lamp*
B *condenser lens*
C *red filter*
E *pentaprism*
F *beam-splitting
   prism*
G *prism*
H *objective lenses*
K *second beam-
   splitting prism*
L & M *photo-cells*
N *moving-coil unit*
*(By courtesy of
E. Leitz, Wetzlar)*

width. Analogue signals corresponding to half the scale division width are thus obtained, and those signals can be divided electronically into 200 parts.

To ensure that the light from the lamp has a constant value it is monitored by a third photo-cell, which operates the gain control of an amplifier in the lamp circuit, increasing the supply to compensate for progressive darkening of the bulb by evaporation from the filament. The prism G is connected to a lever forming part of the moving-coil unit N, so that it can be adjusted to move the scale images for setting zero points.

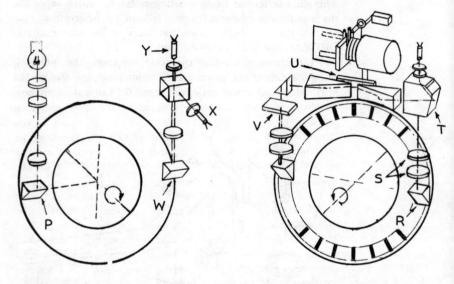

Figure 2.5.
A Leitz incremental angle transducer:

P & R prisms
S lenses
T pentaprism
U mirror
V Wollaston prism
X & Y photo-cells
(By courtesy of E. Leitz, Wetzlar)

The Leitz incremental angle transducer is shown in Figure 2.5., the two drawings showing opposite sides of the glass disc with the scale divisions. These can be arranged to produce either 21,600 or 25,920 signals per disc revolution, to divide the circle into intervals of 1′ of arc or 50″ of arc as desired. Light from the lamp is directed towards the scale, to illuminate up to 200 divisions, and the images of these divisions are directed by the prisms P and R, through lenses S and pentaprism T, to a mirror U, which can be angularly adjusted by means of the moving-coil unit above. From this mirror the images are directed to the Wollaston prism V, whereby the polarized phase-displaced images are produced and re-formed on the surface of the scale as before.

The two images are picked up by the prism W, on the other side of the scale disc, and are passed to a second Wollaston prism which distinguishes the direction of polarization of the light and passes the two signals to the photo-cells X and Y. A maximum of 6,400 scale divisions on the glass disc can be resolved with good contrast by the optical system, and the two phases produced for each division by the contra-rotation of the disc and the images provide for the generation of 12,800 complete signals per disc revolution. These signals can be divided into 51,200 angular increments, corresponding to a shaft movement of 25' of arc, by electronic means, with an error of ±0.5' of arc.

The transducers are unaffected by temperature variations between 10 and 45°C, by vibration or by shocks up to 6.5g. The glass scales measure 20 mm x 30 mm in section and are made in lengths of 1,200 mm, being joined for longer distances.

OPTICAL PROBE IN METROLOGY

For the inspection of delicate components, the combination of an optical instrument with a photo-electric sensor can provide a good solution. Alternatively, a closed-loop system can be introduced which couples the sensor to a measuring transducer. The operator is then only required to set up the equipment to fairly coarse limits, thereby eliminating the possibility of human error. Special optical systems with or without photo-electric sensors can prove satisfactory for the location of inaccessible features.

A notable advance made in this field has been the addition of simple types of photo-electric detectors to microscope systems. Not only does this obviate operator fatigue when making repetitive settings in line graduation of scales, but setting acuity over visually operated microscopes is also improved. The micrometer microscope, which has a built-in measuring transducer, is a good illustration. Instruments designed by Hilger and Watts Ltd are of the photo-electric detector system and are based on a moving-iron vibrator operating at 50 or 60 c/s; these are reliable over long periods and hold the centre of vibration with great accuracy.

A use of the photo-electric micrometer microscope is the comparison of strained specimens. An unstrained sample, having two lines engraved on a polished surface at a separation of, say, 10 mm, is secured in a work holder, while a second sample with a known strain is also secured. The separation of the holders is then adjusted to permit line images from each sample to be superimposed via beam-dividing prisms. The work holders are then displaced by means of a micrometer-controlled stage to the nominal line pitch of the unstrained sample. The microscope micrometer can now be used to determine the difference in pitch to an accuracy better than 0.5 μm.

SURFACE LOCATION

A simple method of non-contact surface location is to use the critical depth of focus of a microscope objective, but an improvement in convenience and accuracy can be obtained by using an auto-reflecting microscope. A target pattern is projected along the optical axis of the microscope and is seen in sharp focus on the surface of the object. Hilger and Watts Ltd have re-arranged the optical lay-out of this type of microscope and have added a photo-electric detector system as shown in Figure 2.6.

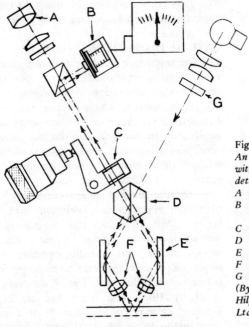

Figure 2.6.
An optical system with a photo-electric detector:
A   eyepiece
B   photo-electric detector
C   refractor block
D   beam splitter
E   mirrors
F   objectives
G   target graticule
(By courtesy of Hilger and Watts Ltd.)

The illuminated graticule pattern is projected through a semi-reflector and two separate objectives onto the workpiece. Images are reflected through the optical system and are viewed by an eyepiece A and a photo-electric detector B. There is a refractor block at C, a beam splitter at D and mirrors at E, these leading to the objectives F. The target graticule is shown at G. Displacement of the work surface in depth causes the images to move across the field of view. Re-alignment of the images with respect to the photo-electric detector is achieved by tilting a refractor block controlled by a micrometer screw until a null reading is given on the meter. Should the surface be tilted, a double image will be formed, but the detector will give an unchanged reading.

The instrument can be applied either as a fiducial indicator or as a measuring instrument; for this, displacements of the image can be related to linear distances on the workpiece, in which case the instrument permits direct reading to 0.5 μm over a range of 0.25 mm. The applications include determination of the separation of encapsulated components, a checking on the flatness of narrow annuli with reference to an optical flat and measurement of hemispheres for gas bearings (where ceramic hemispheres must be checked for height relation to base which is critical). The instrument will operate on materials of various reflectivities from coloured paper to mirror surfaces.

ELECTRONIC GAUGING

The Wayne-Kerr Company Ltd, Surbiton, have developed transducer equipment using non-contacting sensors or probes to cover many engineering problems of precision measurement. The operation relies on the electrical capacitance between the sensor and test surface, and one, two or six channels can be provided. Figure 2.7(a). shows a diagram that indicates checking for displacement, expansion and dilation. These three parameters are a result of movement, either linear or radial. In all cases, if we assume that the

Figure 2.7.
Diagrams showing electronic gauging systems (By courtesy of Wayne-Kerr Ltd.)

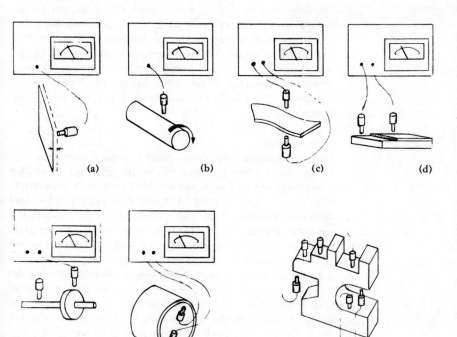

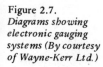

body is electrically conducting, the parameter can be seen as a change in the position of the surface relative to a sensor and causes a response on the indicating meter. A variant relying on the same principle is pressure measurement using a sensor to detect changes in the position of a diaphragm subjected to pressure changes.

Roundness and clearance

This is indicated in Figure 2.7(b). It can be established quickly by using one sensor and by rotating the object. Any out-of-roundness will be shown as a change in the meter reading; the location of the irregularity is shown as lying on the near radius of the shaft in line with the sensor. By allowing the sensor tip to lie beneath the surface of the shaft bearing by a pre-determined amount, working clearances can be monitored. If necessary, a number of sensors can be located at various positions around or along a shaft, and the clearance at each point can be checked for all shaft positions.

Strip thickness

A pair of sensors can be positioned one either side of a metal strip — static or moving — to provide a continuous measure of thickness which does not depend on the exact location of the strip (Figure 2.7(c)). For a given thickness, movement towards one sensor will produce a smaller clearance, but there will be a corresponding increase in clearance between the strip and the other sensor. The instrument will take the sum of the two readings, which will cancel out changes due to strip movement. However, if thickness changes, exact cancellation will not occur and the resultant reading is a measure of the change.

Film coatings

Many non-conducting films possess a permittivity (i.e. an ability to concentrate electric fields) differing from the permittivity of air by an appreciable amount. In such cases, two sensors can be used to derive a measure of the film thickness provided that the permittivity itself is sensibly constant (Figure 2.7(d)). The first sensor is located opposite an uncoated portion of the object. The second one is located at the same distance from the base material but will produce a different reading owing to the presence of the film coating. It is essentially a differential type of measurement, so that bodily movement of the base material affects both sensors, whilst variation in the film thickness affects only the right-hand sensor. This type of measurement is not linear, but thickness values can be easily calculated.

Eccentricity

This is essentially a differential measurement, one sensor monitoring the shaft position which will provide a changing datum should the shaft be warped (Figure 2.7(e)). The second sensor responds to the combined eccentricities due to the shaft and the disc. The measuring instruments can be arranged to read the differences between these two quantities.

Bore readings

Two sensors mounted back-to-back can be inserted into tubing and their signals combined to give a direct reading of the bore (Figure 2.7(f)). By rotating the sensor assembly at a given distance along the tube, any ovality can be detected. The sensors can be incorporated into a cylindrical construction with a loose fit in the bore and centralized by air jets. Such a set-up has been used for evaluating the bores carrying control rods in nuclear reactors, where physical contact can cause contamination.

Inspection and monitoring

The equipment is applicable for rapid checking of dimensions as shown in Figure 2.7(g) and for monitoring machining operations. The sensors can be situated remotely from the measuring instrument, enabling many parameters to be checked in locations difficult for an operator to function.

INSTRUMENTS

On the Wayne-Kerr transducer equipment an outlet is provided for connection to recorders, oscilloscopes or other monitoring devices. Thus any vibration of the test object can be examined. This is valuable because the response is flat to within 0.5 dB up to 3 kHz. The two-channel instrument operates with two sensors, and in addition to a reading value A or B it can be switched for direct read-out of $A + B$ or $A - B$. The six-channel instrument reads $A + B$, $A - B$, $C + D$, $C - D$, $E + F$ or $E - F$. Five standard sensors are available; these are rugged, have no moving parts and can operate with connecting leads up to 10 m in length. Their performance is unaffected by magnetic fields, while the small signal voltage on the end face is harmless to operators. The basis of sensor response is the electrical capacitance existing between the centre electrode and the grounded conducting surface.

For measurement circuits the capacitance of the sensor completes a negative-feedback loop over a high-gain amplifier which has a reference input from a 16 kHz oscillator, the output amplitude being directly proportional to the sensor-structure separation. Thus, if the structure is vibrating, the

Figure 2.8.
A two-axis point-to-point positioning system showing the arrangement of cylinders and measuring bars for table positioning (By courtesy of Plessey Ltd.)

output will be amplitude modulated. The signal is applied to rectifying and filtering circuits which produce d.c. voltage. This is fed, via switching circuits and sum/difference amplifiers, to the meter.

There are built-in limit circuits giving outputs for automatic rejection of out-of-tolerance components, while a further option is the provision of combinations of measurement channels to provide reading of, say, $A + B - 2C$; this has been used for checking the straightness of a shaft.

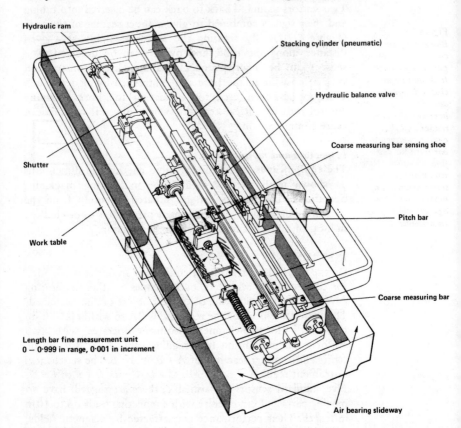

Hydraulic ram

Stacking cylinder (pneumatic)

Hydraulic balance valve

Coarse measuring bar sensing shoe

Shutter

Pitch bar

Work table

Coarse measuring bar

Length bar fine measurement unit
0 − 0·999 in range, 0·001 in increment

Air bearing slideway

Fluidic Numerical Control (NC) System

A low-cost two-axis point-to-point positioning system has been designed by the Plessey Co. Ltd. The system which comprises a co-ordinate table and control unit is shown in Figure 2.8. which shows the arrangement of the cylinders and the measuring bars for the table positioning. The system functions by a combination of pneumatic and hydraulic power and has been devised for use with vertical drilling machines, but it can be incorporated in new machines or can replace the table

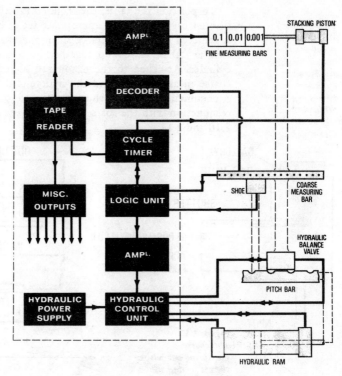

Figure 2.9.
Block diagram of control elements of positioning system. It should be noted that a decimal decoder is necessary between the tape reader and the coarse measuring bar but that the fine measuring bars, which are themselves arranged in binary form, can operate direct (via an amplifier)

of machines already in use. There are no electronics associated with the unit, for control is accomplished by means of a low-pressure pneumatic block paper tape reader, decoder, cycle timer and turbulance amplifier logic elements. These are coupled through high-gain pneumatic relays to conventional air-piloted hydraulic and pneumatic control valves; Figure 2.9. shows a block diagram of the system where hydraulic rams, operating at 10 kg/cm^2 provide the table movement.

The program is punched on standard eight-hole paper tape. On command, the fluid cycle timer releases the table clamps, admits air to the table air bearing and releases the tape clamp. The paper tape advances on the information block, and the low-pressure signals from the tape reader are amplified and are used to set the fine and coarse measuring systems.

The coarse measuring bar postions the table to the nearest inch, and fine positioning is determined by a set of measuring bars, arranged in binary form, which can be adjusted over 0.999 in. by increments of 0.001 in. These operate in conjunction with a pitch bar and hydraulic balance valve, the arrangement shown being similar for both the table and traverse slide.

To position the table, the required increment of an inch is first set by selecting the appropriate set of bars; this selection is achieved simply by a set of small pneumatic cylinders. The bars are then closed up together by a pneumatic stacking cylinder. Attached to the piston rod of this cylinder are the coarse measuring bar, and the hydraulic balance valve, while a pneumatic shoe and shutter attached to the table act in conjunction with the coarse measuring bar as shown in Figures 2.10 and 2.11.

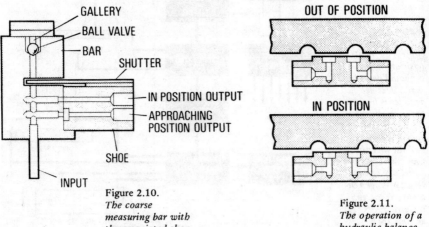

Figure 2.10.
*The coarse
measuring bar with
the associated shoe
and shutter*

Figure 2.11.
*The operation of a
hydraulic balance
valve*

A signal is fed to the appropriate inch position input on the coarse measuring bar, and the cycle timer is arrested at this point. The shutter has its leading edge in line with the 'in position' output hole on the shoe and determines, in the first instance, in which direction the table should traverse by stopping all signals which are negative to the present position. In other words, if an output signal is obtained from the gallery, the table will advance; if there is no output signal, the table will retreat. The presence or absence of a signal determines the position of a three-position hydraulic valve which puts full pressure on one side of the hydraulic piston and exhausts the other side.

The piston moves the table at a fast traverse rate until a signal is obtained from the 'approaching position' output on the pneumatic shoe. The fast traverse valve closes, and hydraulic piston control is transferred to the balance valve, working in conjunction with the fine measuring pitch bar. A hydraulic bridge is formed which centres the table accurately on the required notch on the pitch bar as shown in Figure 2.12.

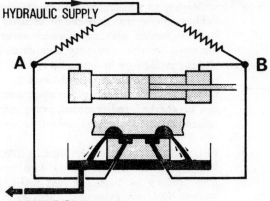

Figure 2.12.
A schematic arrangement of the hydraulic bridge

The coarse-position shoe transmits a signal from the 'in position' output, and the cycle timer is unlatched. After a short delay to allow the table to come to rest, the timer completes the cycle by cutting off the air supply to the table air bearing, clamping the table in position, damping the hydraulic pressure and signalling the tool to commence the machining operation.

There are four channels on the tape for miscellaneous functions; the positioning accuracy of the table is 0.05 mm, and the repeat accuracy is ± 0.025 mm. The fast traverse rate is 6,000 mm/min, and the table load capacity is 200 kg. Figure 2.13 shows a graph indicating the variation in pressure on each side of the hydraulic piston compared with the displacement of the machine table.

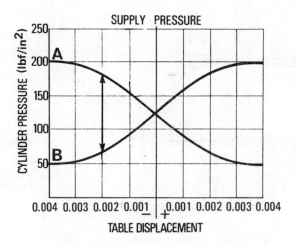

Figure 2.13.
A graph of the pressure variation on the piston with table displacement

Machines with accurate table-positioning systems find useful applications in that they can be used for machining purposes such as boring operations where accurate hole spacing is essential. In this way they can eliminate the use of jigs, thereby saving time in setting up and also the cost of expensive jigs. If a co-ordinate table is used in conjunction with an indicating probe on a ram, errors in centre distances of holes can be checked and recorded, in a similar way to those on a co-ordinate measuring machine but only two axes are needed.

PEL SYSTEM OF PNEUMATIC ELECTRIC SIZING

The system makes use of standard modules and has been originated by the Swiss company SAIA (Landis and Gyr Ltd, Acton). The basic element is a gauging switch which combines both pneumatic and electrical principles, and which is employed in conjunction with one or more detector nozzles. Comparison with the Wheatstone bridge circuit will enable the operating principles of the system to be understood. Figure 2.14(a) shows a pneumatic bridge circuit in comparison with the Wheatstone bridge of Figure 2.14(b). The pneumatic bridge acts in a similar manner to the electrical bridge and has flow resistors in place of electrical ones; the pneumatic system is largely independent of supply pressure p. Piston K (Figure 2.15) corresponds to the galvanometer G in the electrical circuit, and the piston responds to very small changes in the air flow from the detector nozzle R_x. The orifice R_v is adjustable and corresponds to the variable resistor R_v in the electrical bridge.

Figure 2.14.
The Pel system of pneumatic electric gauging (By courtesy of Landis and Gyr Ltd.)

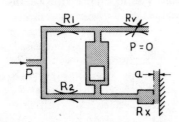

(a)

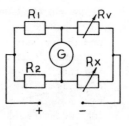

(b)

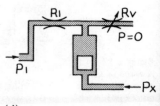

(c)

(d)

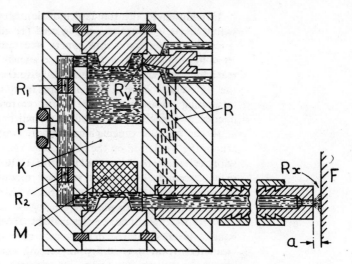

Figure 2.15.
*The measuring head
of a pneumatic
measuring system*

An electrical bridge is balanced before use, to provide zero current to a galvanometer G, so that, if the resistor R_x is of a temperature-sensitive type, variations in temperature cause a change in resistance which, in turn, upsets the balance of the bridge. A current then flows through the galvanometer G to give an indication of the temperature change. The action of the pneumatic bridge in the Pel system operates in a similar manner. If it is desired to measure the clearance a (Figure 2.15) between the detector nozzle and the surface F, the nozzle reacts in a similar manner to that of the electrical resistors. As the nozzle is moved towards the surface, the flow resistance and the pressure in that branch of the circuit are increased. Consequently there is a pressure difference between the R_v and the R_x branches of the circuit, to which the piston K responds and which is thus caused to rise. A pressure at R_x that is only 30 mm WG higher than that at R_v is sufficient for this purpose. The piston incorporates a permanent magnet M, the field of which actuates the reed switch R so that a current flows in an electrical circuit. The differential or the movement that the surface F must make relative to nozzle R_x to produce an on-off cycle is 2 μm under optimum conditions.

It is possible to connect two or more gauging switches in series so that each operates when there is a pre-determined clearance between the detector nozzle and the surface F. Thus, by assembling the modules in the required sequence, components may be sorted into tolerance groups.

In addition to the standard module, there is a differential switch and a pressure gauging switch. The differential switch is shown schematically in Figure 2.14(c), and it incorporates an additional connection for a second nozzle R_x. This arrangement increases the sensitivity of the system which responds to the difference between the dimensions a_1 and a_2, thereby facilitating measurement of tapers, matching and similar operations for which the nozzles are used in oppostion.

In the pressure gauging switch (Figure 2.14(d)), the lower arm of the T-junction on the inlet side of the circuit is omitted, with the result that the system is extremely sensitive to changes in air pressure. The supply pressure p_1 may be varied, thus allowing control by means of pressure surges or accurate monitoring of static air pressure.

Forty-six types of nozzles are available. These are as follows: control nozzles for the detection of movements of parts at clearances from 0 to 0.5 mm; sensing nozzles for gauging clearances between 0.03 and 0.25 mm; sub-miniature nozzles which enables gauging to be carried out in a small area — up to 20 dimensions/cm^2 can be gauged simultaneously; concentric nozzles which increase the effective range of the system up to 3.5 mm; pin-type nozzles in which a taper pin slides up and down the bore of the nozzle, thus providing a means of extending the measuring range. Pressure gauging switches can be used in conjunction with transmitter and receiver nozzles to form 'detector gates' which operate at distances up to 100 mm and are used for pressure detection and counting applications.

3.
The Laser & Precision Measurement

Introduction

A laser is an oscillator and must be supplied with power from which to select its own characteristic form of energy and to store it, as shown in Figure 3.1. The store is topped up from the power source, thus compensating for the small amounts of energy that are lost by allowing some energy to emerge usefully and by other less useful side effects. The system must be kept well topped up, and too little pumping or excessive extraction of energy will destroy the selection process so that the device ceases to function.

The pump in this example is the electrical drive, and the selection process is performed by the discharge tube which contains helium and neon, and in many respects this is similar to a conventional neon light. Storage is accomplished by bouncing the energy to and fro between carefully aligned mirrors, and extraction is made possible by arranging for one mirror to transmit a small amount of energy. Various types of lasers differ essentially only in their selection and pumping process. Development of the laser made it possible to create clear coherent light, and it is the coherence of the beam from a laser that makes it so significant to the engineer and physicist.

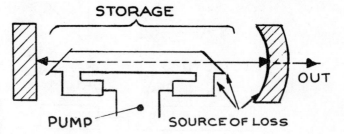

Figure 3.1.
A diagram showing the power supply of a laser

Basic Beam

One effect of coherence is that all the energy appears to come from a very small point, and the simplest of lenses incorporated in the laser will re-focus the beam to a very small point or into a parallel beam. The typical helium-helium laser produces a beam of pure red light, 1 or 2 mm in diameter, with a power of 1 mW. Although the power is small, when it is focused it has an intensity that is several times greater than the brightest conventional lamp. At a distance of several metres, the beam begins to expand at a rate of about 1 mm/m.

The spot of light is readily visible on virtually any surface, and its centre is judged easily or is measured in the most casual fashion to an accuracy of 1 mm over 20 m. Thus with the aid of a few simple accessories, such as graph paper, mirrors and screens with holes, a very useful instrument can be created. The application of lasers to inspection is valuable in that the beam is visible on any screen, it is observed easily from a distance and it can remain in position for reference.

The range over which the basic laser can be used is restricted to about 50 m because of the divergence of the beam. For a beam 25 mm in diameter, the ultimate divergence could be 5 mm in 100 m. By adjustment, the beam can be made parallel for a distance of about 300 m.

Interferometry

The laser is pre-eminent in the field of very accurate measurement of length, e.g. of the order of 0.1 μm in 100 m. It has removed most of the time and skill required previously for making such high-accuracy measurements. The laser design is modified, at the expense of total power, so that a precisely defined single frequency is selected from the coherent beam and is used for interferometric measurement. A representative laser interferometer is shown in Figure 3.2.

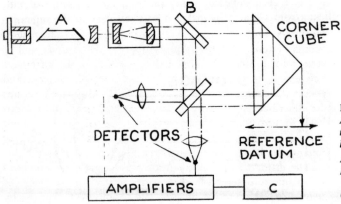

Figure 3.2.
A diagram showing the functioning of a laser inferometer:

A   *He-Ne laser*
B   *reference beam-splitter*
C   *counter*

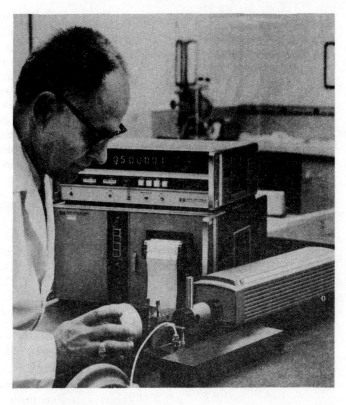

Figure 3.3.
An application of
the laser in checking
micrometer heads
(By courtesy of
Hewlett-Packard
Ltd.)

HELIUM-NEON LASER

A low-power helium-neon laser A emits a coherent light beam composed of two slightly different optical frequencies f_1 and f_2 with opposite circular polarizations. After conversion to orthogonal linear polarizations the beam is expanded and collimated; then it is directed to the reference beam splitter B where a small portion of both frequencies is split off. This portion of the beam is used both to generate a reference frequency and to provide an error signal to a laser cavity tuning system. The difference in the d.c. levels of f_1 and f_2 is used for tuning, while the a.c. component of the difference between f_1 and f_2 (about 1.8 MHz) is used for reference and goes to a counter in the laser display C. Alternative set-ups are described later; these are from developments of Hewlett-Packard Ltd, Slough, while an illustration of the equipment is given in Figure 3.3, which shows its application to checking micrometer heads to within millionths of an inch.

The major portion of the beam passes out of the laser head to an interferometer which measures relative displacement of two retro-reflectors by splitting the beam into f_1 and f_2, by directing them to two retro-reflectors and by returning the resultant signals to a photo-detector in the laser head. Relative motion between the retro-reflectors causes a difference in the Doppler shifts in the return frequencies, thus creating a phase difference between the frequency seen by the measurement photo-detector and that seen by the reference one. The difference is monitored by a subtractor and is accumulated in a fringe-count register.

A digital calculator samples the accumulated value every 5 ms and performs a two-stage multiplication, one for refractive index correction and the other for conversion to millimetres. The resulting value updates the display.

Linear Measurement

A feature of the Hewlett-Packard system is that by using optional equipment a wide variety of measurements and inspection can be carried out. Figure 3.4(a) shows the beam exiting from the laser head and being split up at the surface of a polarizing beam splitter with one frequency reflected to the reference cube corner mounted on the housing. The other is transmitted to the measuring retro-reflector. Both frequencies are reflected back along a common axis to the photo-detector block in the laser head, one of which includes a Doppler frequency shift whenever the measurement retro-reflector moves. Since their polarizations are orthogonal to each other, they do not interfere to form fringes until the beam reaches the de-modulating polarizer mounted in front of the photo-detector.

Angular / Flatness

A $45°$ mirror is mounted in place of the reference retro-reflector so that f_1 and f_2 are sent out parallel (Figure 3.4(b)). Angular displacement of the retro-reflector mount causes a differential Doppler shift in the returned frequencies which is not affected by axial displacement. The accumulated fringe counts are proportional to the sine of the angular displacement.

Single-beam Interferometry

A polarizing beam splitter reflects f_2 to the reference retro-reflector and transmits f_1 to the surface whose displacement is being measured (Figure 3.4(c)). Since both beams leaving the beam splitter pass through a quarter-wave plate, the returning polarizations are rotated through $90°$. This causes f_2 to be transmitted and $f_1 \pm \Delta f$ to be reflected so that they are returned co-axially to the laser head by the beam-bending mirror.

Plane Mirror Interferometry

The beam entering the interferometer is split into f_1 and f_2, with f_2 returning to the laser head after retro-reflection by the reference cube corner (Figure 3.4(d)), as in the linear interferometer. f_1 is transmitted out to the plane reflector and is

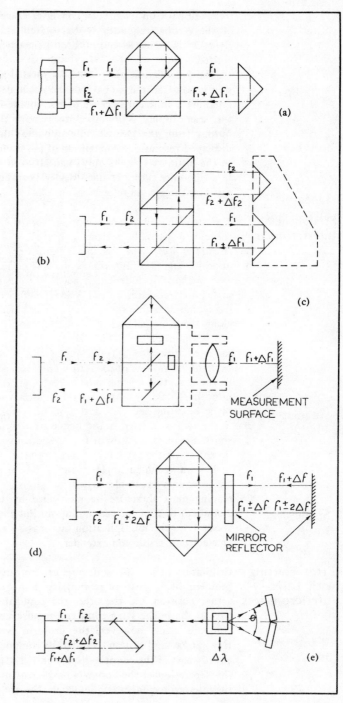

Figure 3.4.
Diagrams showing the variety of applications in measuring and inspection by lasers

reflected back on itself. The converter causes the polarization of the return frequency to be rotated through $90°$ so that $f_1 \pm \Delta f$ is reflected out a second time where it is Doppler shifted again.

The polarization of $f_1 \pm 2\Delta f$ is rotated again through $90°$, so it is now transmitted back to the photo-detector. Resolution doubling is inherent because of the double Doppler shift, but the laser display is modified to remove the doubler in the output from the Doppler pre-amplifier, thus correcting the displayed resolution. Any tilting of the plane reflector relative to the beam axis results only in an offset of the return and not in a tilt, since tilting of the reflected beams is exactly compensated by the second reflection.

Straightness Interferometry

The two-frequency beam exiting from the laser head passes through a partially reflecting mirror to a Wollaston prism (Figure 3.4(e)). The two polarizations are refracted at equal opposite angles about the centre-line, causing f_1 and f_2 to exit at a small included angle matched to that of the Wollaston prism interferometer. f_1 and f_2 are therefore reflected normally to recombine within the prism. The combined beam is returned along the same path as the exit beam to the partial mirror in the beam displacer. The majority of the returned signal is reflected down to a 100% mirror which deflects the return beam back into one of the return apertures of the laser head.

Axial motion of the prism creates an equal Doppler shift in both f_1 and f_2. Any lateral motion with respect to the plane bisecting the two mirror planes results in a difference in the two path lengths and hence a difference in the Doppler shifts. Small pitch, yaw or roll motions of the interferometer do not affect the accuracy of measurement.

For an included angle θ and a lateral translation x, the fringe counts accumulated will be given by $2x \sin(\theta/2)$. The reading must therefore be multiplied by the reciprocal of $2 \sin(\theta/2)$ to give correct resolution. For the value of θ used, about $1.5°$, the multiplication factor is 36 and is provided by an electronic resolution extender.

Error Plotting with Laser Interferometers

Calibration of measuring devices can be carried out, the basic problem being that of comparing two quantities, one the actual position and the other the command position. These have to be in the same form and require a separate device to compute error; however, the problem is overcome in the Hewlett-Packard system by using the computational capability, already part of the counter-display unit. The technique involves the separation of the command-and-error measurements from the interferometer reading in which they are combined, i.e.

interferometer reading = command position + error

As an example, let us suppose that the machine has been positioned to 4.000000 in. and that the true position as measured by the interferometer is 4.00012 in. The reading is split into two parts: the left-hand part is taken as the command position; the other part 12 is taken as the error in ten-millionths of an inch. An analogue voltage corresponding to each part is transmitted to the recorder which moves to a point at the X and Y co-ordinates of 4.000 and + 0.00012 in. respectively.

If the reading is indicative of a negative error, the right-hand digits are then incremented by one unit so that the nominal position will not be plotted one unit too low.

Applications

There are three modes of operation for the error-plotting device: these are manual, continuous and step-by-step.

MANUAL

The simplest mode is the manual and can be used for the calibration of micrometers, optical scales, manually positioned machine tools and other distance-reading devices. As Figure 3.5 shows, the only electrical connections are from the display unit A to the X-Y recorder. A bench micrometer is shown as an example.

Once the laser beam has been aligned with the micrometer axis this is set to zero. The operator re-sets the interferometer and adjusts the X-Y recorder for zero error. After moving the reflector to the first calibration point, 0.25 mm for example, the operator presses the manual plot switch B. This causes an error computation to be made on the measurement being displayed by the interferometer, and a point is plotted by the recorder corresponding to this error and the command position of 0.25 mm. The cycle is then repeated for the desired calibration range.

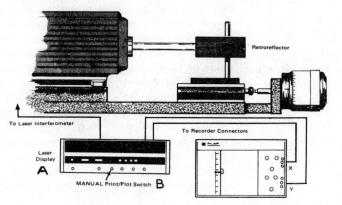

Figure 3.5.
A diagram showing the set-up for error plotting of a bench micrometer.

CONTINUOUS

The most efficient way to calibrate NC machine tools and other servo-positioned systems is continuous or 'on-the-fly' error plotting. The table or saddle is traversed along the calibration axis and controls the plotting rate. All that is needed is a pulse or contact closure at each desired calibration point, the signal being tapped off from a shaft encoder or the digital read-out circuitry. Other methods of generating command trigger signals include microswitch-gear and slotted disc-lamp photo-cell systems.

Figure 3.6 shows the connections for continuous calibration of an NC milling machine axis. The maximum plotting rate, and hence the traverse feed rate, is dependent upon the magnitude of the error signals and of the plotting accuracy required. It is generally between 10 and 25 points/s.

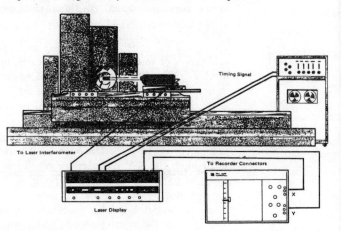

Figure 3.6.
A view of the connections for continuous calibration of a milling table

STEP-BY-STEP

This method is also applicable to NC machine tools but is not continuous. The error-plotting rate is controlled by a timer built into the laser display and is varied by a front panel switch. At each timed point a brief contact closure is available at rear panel connectors. If this circuit can be used to advance a table or saddle by a constant increment, then a stepping mode of calibration is possible. A machine tool can be programmed for a step-by-step error plot by making a command tape to advance the table by the calibration increment on receipt of a timed pulse at the tape advance switch. Since the machine tool must stop at each calibration point, the time to check each axis is greater than for continuous plotting.

An interesting calibration is that of checking the pitch and yaw of the table as it moves along the bed. This can be done by using two lasers, each aimed at measuring the position of a different point on the moving table. Their error outputs are connected so that the difference between the two readings, corresponding to yaw or pitch, is plotted against the travel of the table. It is often important to know how the dimensions change when the machine spindle is running, and by use of the laser it is possible to measure and record the axial growth in a spindle due to the generation of heat within the bearings. The cube corner reflector is mounted in a chuck by the stub shaft provided, and the laser beam is directed into the reflector from the interferometer mounted on the table or carriage. In this application the 'error' output is connected to a strip chart recorder to show growth against time.

Figure 3.7 shows the application of a laser measurement system to the traverse motions on a horizontal boring machine: the three applications are to measure (1) the vertical traverse of the saddle on the column, (2) the longitudinal movement of the table on the bed and (3) the transverse traverse of the table. The advantages include the feature that the design uses only a handful of small optical and electronic components and a single remote laser source, all easy to install. There are no parts subject to wear; nor is periodic calibration of the transducer required for the wavelength of a laser beam is a length standard in its own right. Pitch and yaw can be measured as well as position, and the system is capable of compensating for temperature changes in the machine parts and workpieces.

The technical specification of the laser transducer is as follows: resolution, 0.15 μm; accuracy 0.5 μm/m; maximum slew rate, 0.3 m/s; number of axes, 1 to 8; range, 60 m sum of axes.

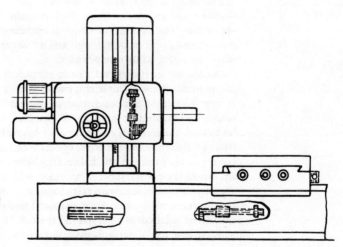

Figure 3.7.
The application of a laser measurement system to a NC boring machine.

4.
Electronic Inspection & Measuring Machines

Introduction

The development of NC or computer-controlled manufacture has now indicated the need for similar methods of inspection instead of traditional methods which can take a hundred times as long as the length of manufacture of the same component. Two categories of inspection are desirable: point inspection and continuous inspection. The first one involves the programming of a light accurate machine with an inspection head to move it to pre-determined points and to record any error from the true position. Continuous inspection is desirable for workpieces where the whole surface is important, e.g. aerofoils, turbine blades and templates.

Ferranti Inspection and Measuring Machines

A range is available from the Mercury machine (with a measuring head traversing in the three axes X, Y and Z of 508, 406 and 203 mm) to a large four-column Saturn machine (with a measuring capacity of 2,000, 1,250 and 1,000 mm respectively). Typical of the Ferranti design is the Hydro-cord which is an advanced three-axis measuring machine using 'constraint' sensing and embodying new developments in hydrostatics and kinematics. High-precision measurement is obtainable, the resolution (digit size) being 0.001 mm with a repeatability of ± 0.002 mm and an accuracy over the full movement of ± 0.007 mm per axis.

As shown in Figure 4.1 the construction is a fixed-gantry moving-table design. This arrangement gives maximum stiffness in the two most important axes X and Y without compounding. Maximum stability is obtained by the one-piece fabricated torsion-box construction. A probe column on the front of the X carriage is counter-balanced with an adjustment control which enables the balance to be set to suit the probe tip in use. The work table is provided with a slew control for aligning the workpiece with respect to the X and Y axes.

The work table, the X carriage and the probe column are supported on hydrostatic pads (Figure 4.2). Oil between the pads A and the slideways gives a damped friction-free characteristic, the pads being fed with oil at a controlled pressure and temperature. Similarly, an automatic control valve compensates for loads placed on the work table. Because of the friction-free design accurate measurement can be made by

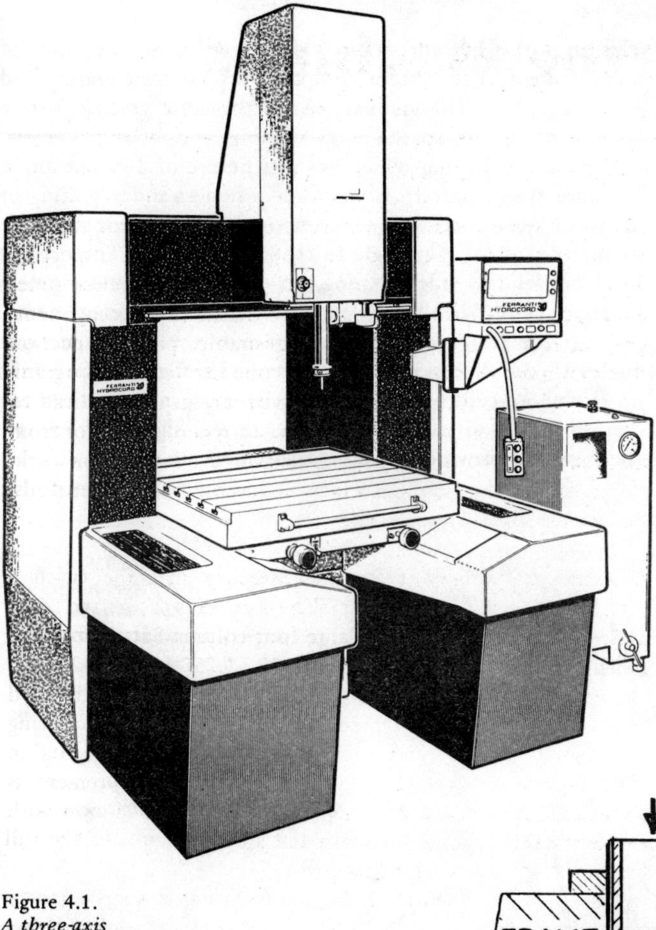

Figure 4.1.
A three-axis measuring machine using 'constraint' sensing (By courtesy of Ferranti Ltd.)

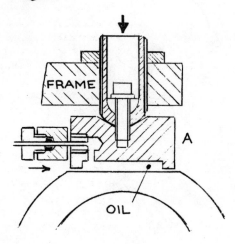

Figure 4.2.
Hydraulic pads (A) supporting work table, carriage and probe column

constraint sensing which is a measuring technique where the probe is brought into contact with a selected feature of the workpiece and is allowed to settle naturally. No sensitive indicator of contact is used, the positioning accuracy being dependent on the freedom of movement of the carriages.

Moiré Fringe System of Measurement

This uses the Ferranti moiré fringe system on each axis of measurement. The system comprises a glass scale grating and a reading head. The graduations on the scale grating form a pattern of closely spaced lines running at right angles to the length of the grating which has a structure of 100 lines/mm. A moiré fringe pattern is produced when an index grating of identical line structure is superimposed with the line structure of one set at a slight angle to that of the other. The reading head houses the index grating, an exciter lamp and a group of photo-cells. When the index grating is moved along the scale grating, the fringe pattern travels across the grating at right angles to the direction of movement. The light pattern falls on the photo-cells which converts the changes in light density (sinusoidal in distribution) into electrical signals which are then used to provide precise measurement of travel.

Figure 4.3.
An optical system using phase trans-mission gratings:
1   *exciter lamp*
2   *collimating lens*
3   *index grating*
4   *scale grating*
5   *mask*
6   *four-section lens*
7   *output slit*
8   *photo-cells*

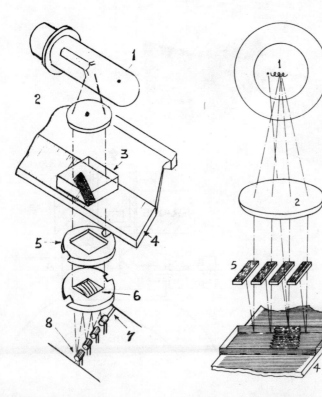

Figure 4.4.
A diagram showing the optical arrange-ment with reflecting gratings:
1   *filament*
2   *collimating lens*
3   *index grating*
4   *scale grating*
5   *strip-silicon photo-cells*

Various optical arrangements are possible; Figure 4.3 shows an optical system using phase transmission gratings. Figure 4.4 shows the optical arrangement with reflecting gratings. A third arrangement, Figure 4.5, is used with line and space transmission gratings. The phase type of grating is normally produced on selected glass blanks: the metal reflecting gratings cater for travel lengths greater than 1,800 mm, while the line and space gratings include the widely used 40 lines/mm variety which are reproduced photographically on glass.

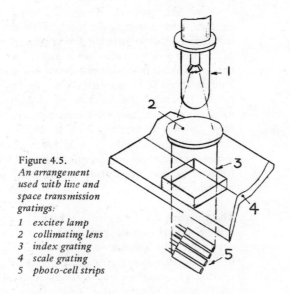

Figure 4.5.
An arrangement used with line and space transmission gratings:

1 exciter lamp
2 collimating lens
3 index grating
4 scale grating
5 photo-cell strips

The accuracy of the measuring system is assessed by considering each separate source of error and by computing the root mean square. Since each measurement involves two settings, it is necessary to add in twice the square of each error. The main error sources are as follows.

(1) The grating error E_G can be taken as the calibration specification limit for the length and type of grating in use which is given for 99% confidence level.

(2) The digital error E_D is taken at a 2σ value of 0.7 digit.

(3) The analogue subdivision error E_A arises from tolerances in the d.c. level of the signals from the reading head and in the analogue-digital converter stages which follow. Its value for glass transmission gratings is taken as 0.05 of a grating pitch and 0.08 for steel reflecting gratings.

The overall system error is given by

$$E_S = (2E_G{}^2 + 2E_D{}^2 + 2E_A{}^2)^{1/2}$$

For a subdivision of 10 digits per grating pitch, $E_A = 0.7E_D$ and $1.1E_D$ for glass and steel gratings respectively, giving the simplified formulae,

$$E_S = (2E_G{}^2 + 3E_D{}^2)^{1/2} \text{ for glass}$$

and

$$E_S = (2E_G{}^2 + 4.6E_D{}^2)^{1/2} \text{ for steel.}$$

If mechanical errors are included, the error E_M can be combined with the system error by a further root-mean-square determination to give the overall installation error,

$$E_1 = (E_S{}^2 + E_M{}^2)^{1/2}$$

A modern computer numerical control (CNC) system may be used with a measuring machine to undertake the many calculations that are needed during measurement of a complex part. Such a system can use a part of its memory to store details of small inaccuracies in the measuring machine and to apply a correction automatically to the various measurements taken. However, the machine must initially be calibrated, usually against a laser interferometer, and error information must be stored together with a simple programme for its continuous application. If wear takes place in the machine during its life, re-calibration can provide new error information for storage and application in the same way as when the machine was new.

Electronic Counter

The counting system is identical for each axis of measurement. The electrical signals generated in the reading head are converted into short-duration pulses and are registered in a bi-directional counter. Usually a divide-by-ten system in the counter produces ten pulses for each output cycle from the photo-cells, i.e. ten pulses result from the passage of one moiré fringe across the aperture of the reading head. Each pulse represents an increment of displacement one-tenth of the distance between each line on the grating. In a metric system (a grating line structure of 100 lines/mm) the unit of measurement is 0.001 mm. A number of counting stages are used (one for each numerical indicator) with overflow from one stage being passed on to the next as in conventional counting. The counter keeps a continuous record of the position of a carriage, adding and subtracting digits.

Computer-aided Inspection

Ferranti inspection machines can be integrated to a general-purpose computer and teleprinter. Examples of the basic programmes are given with diagrams in Figure 4.6(a) to (e).

Programme 1: Automatic alignment computation
This programme calculates the misalignment angle between the axis of the inspection machine and the datum axis of the

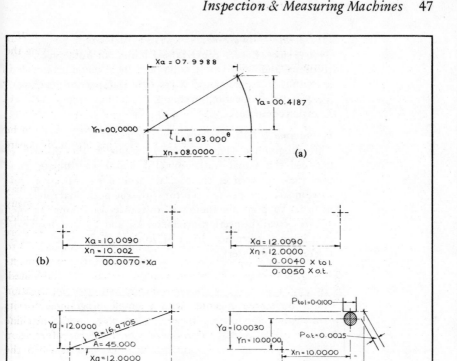

Figure 4.6.
*Basic programmes
which can be
integrated using
computer and
teleprinter*

workpiece. From then on, the computer automatically compensates for this when calculating workpiece dimensions (Figure 4.6(a)). X_n and Y_n are nominal (correct alignment); A is angle (computer-correction factor); X_a and Y_a are actual (inspection machine-correction factor).

Programme 2: Difference computation
This programme (Figure 4.6(b)) calculates the difference between nominal and actual dimensions. The sequence number and nominal dimensions are transferred from pre-punched paper tape and stored for each axis of measurement. X_a is actual (inspection machine); X_n is nominal; X_d is the difference (computer). This is repeated for Y and/or Z.

Programme 3: Out-of-tolerance computation
This (Figure 4.6(c)) causes the printer to record sequence number, axis of measurement, plus-and-minus tolerances, nominal and actual co-ordinates and out-of-tolerance. No figures appear if the actual dimension is within tolerance. X_a is actual (inspection machine); X_n is nominal; X_{tol} is tolerance; $X_{o.t.}$ is out-of-tolerance (computer). This is repeated for Y and/or Z.

Programme 4: Co-ordinate conversion (cartesian to polar)
This (Figure 4.6(d)) converts and prints out both types. The tolerance and deviation of radius and angle would be included as output from computer. X_a is actual (inspection machine); Y_a is actual (inspection machine); R is radius (computer); A is angle (computer).

Programme 5: Polar deviation (true position)
This programme (Figure 4.6(e)) calculates the difference between the actual and nominal X and Y co-ordinates and uses these differences to calculate true positional error and to compare true positional error with true positional tolerance in order to print out the out-of-tolerance conditions. X_a and Y_a are actual (inspection machine); X_n and Y_n are nominal; P_{tol} is polar tolerance (band or zone); $P_{\overrightarrow{o.t.}}$ is polar out-of-tolerance (computer).

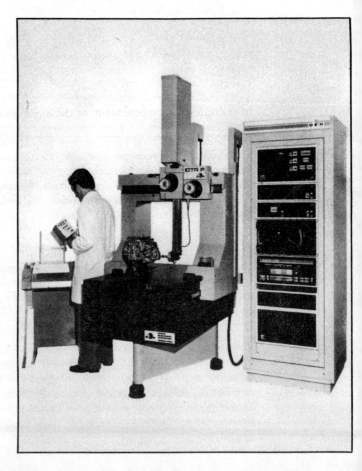

Figure 4.7.
Checking the gear box of a Volkswagen car (By courtesy of DEA, Turin)

Continental Developments

Prominent in the field of high-precision measuring machines is the firm DEA of Turin, Italy. Several sizes of machines are made and each of them for a particular group of applications. The IOTA-P is the smallest and latest development with a capacity of 760 mm × 500 mm × 400 mm traverses; it consists of a cast iron bridge structure which carries the vertical ram and which itself runs on air bearings on a high-precision granite table. The machine embodies binary coded optical scales which are scanned by light sensitive encoders. The air bearings enable a comparatively heavy machine to be moved with ease, while careful design has been applied to the seals of the bearings to prevent appreciable air flow which would possibly chill the granite table if the bridge was left in one position for a period and would cause non-linearity. The machine is shown in Figure 4.7, the operation being that of checking the gear box of a Volkswagen car.

Larger machines are to be found in the Beta and Gamma ranges, with working traverses up to 9,970 mm × 1,600 mm × 1,000 mm. The Beta machines are available with a resolution of either 0.05 or 0.01 mm, and the Gamma has a resolution of 0.005 mm. Generally, these machines have a base of the tower type with isostatic supports of welded steel and slideways of hardened chromium steel stabilized in liquid nitrogen. There is a patented anti-hysteresis device to provide high repeatability of measurements independent of the direction of movement. Beta machines over 1,800 mm traverse (X axis) do not incorporate an integral table, as it is considered that machine isolation from workpiece weight is a vital factor on larger machines.

Another machine is the Sigma-D which also embodies new developments. Rapid and sensitive adjustment is provided for any deflection caused by workpiece weight, and the main and central carriage weight is supported on a patented system of guideways mounted on pillars (Figure 4.8). Deflection of the guideways is thus prevented, while the machine accuracy remains independent of deflections owing to the position of the two carriages (see Figure 4.13).

Movement of the machine is motorized on all three axes, and rotary air bearings are used. Probes can be indexed with precision between $+X$, $-X$, $+Y$ and $-Y$ directions, thus providing a true three-dimensional approach to workpieces without wasted time in changing probes manually. The slideway displacements on this machine are by precision rack segments which are carried independently of the driving racks, and positional information is provided by electronically scanned optical scales. The measuring strokes are 1,800 mm × 1,000 mm × 700 mm, and a workpiece mass of 3,000 kg can be accommodated..

Figure 4.8.
*A DEA machine
motorized on three
axes with movement
on rotary air
bearings*

OPERATIONAL MODE

The Beta machine is typical of the DEA range in its manner of operation, and the tool-carrying head bears five holes parallel to the X, Y and Z axes, their respective centres being co-incidental with that of the Z-axis ram, Figure 4.9. The five-sided tool holder allows the workpiece to be approached from five sides at one setting, thus eliminating re-setting times and inaccuracies. Also it will be seen that the component to be checked can be mounted on any one of its six sides (on the assumption that it is cubic in shape) where a machine with only one vertical hole in its ram can only accept its workpiece in one plane and can only check one side of it at once.

The position of the tool-carrying head is indicated on three main displays in numerical form. Six digits are employed for each axis, and the position of the decimal point is dependent on whether the machine is displaying imperial or metric dimensions. Each axis read-out can be re-set to zero at any point, and a switch is provided to 'freeze' each read-out, thus providing the facility to insert known constants into each read-out and then to move the toolhead to a particular point without losing the constants already inserted. The operator can thus select his datum point relative to the three axes and can have automatic compensation for either diameter or length of the tool in use as in Figure 4.10.

Figure 4.9.
A Beta measuring machine with a five-sided tool holder

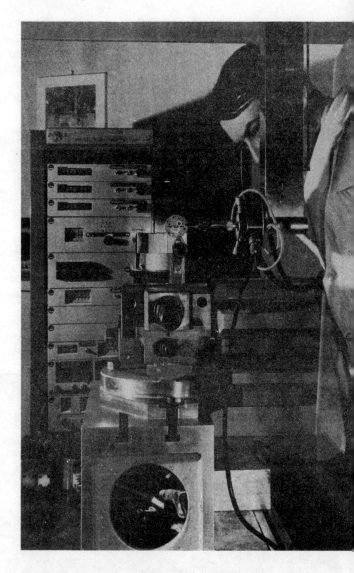

Figure 4.10.
*Tools with
automatic
compensation
for diameter and
length*

The three movements can be separately locked by an electro-magnetic system with fine positioning on each axis carried out by electric motors after the clamping has been effected. This ensures that, once a position has been reached, clamping on that particular axis will not affect that setting. The repeatability is always within the resolution together with an accuracy of positioning within ± 0.02 mm over any 1,000 mm stroke for the largest machine in the range of eighteen machines, with its ram on Z axis at maximum extension.

The Delta and Alpha machines are the largest and most complex of the group. The Delta machine may embody a built-in table and motorized movements and can be controlled from the DEAC 1000 computer over the whole of its movements which are up to 4,700 mm × 2,400 mm × 1,000 mm in the case of the Delta BS. Alpha machines have similar capabilities but without the built-in table, and the largest machine is the Alpha MS with useful traverses of 10,000 mm × 5,500 mm × 2,500 mm. These machines were developed for continuous path scanning of motor car body dies and aircraft parts where speed, accuracy and large capacity are essential. The application of automatic inspection will be discussed in Chapter 11.

An offshoot of these large machines is the DEA Leonardo 02 which as a special machine was made for the Ford Motor Co. of America. The machine is mobile and can be driven around the body-styling plant. It operates from batteries; it can perform continuous path scanning from the control of the computer carried aboard and can produce tapes containing complete dimensional information of a full-size body model. Normally, these models are made from plaster with the consequent danger of breakage if they are moved, but the Leonardo measuring machine eliminates the need to move them.

Integrated Circuits

Although DEA were pioneers in adopting (for industrial electronics) silicon planar transistors, these have now been replaced by integrated circuits which have been developed for data processing. An integrated circuit is a functional block containing several transistors, diodes, resistors and capacitors in an extremely small volume; it gives a performance equivalent to the more cumbersome conventional electronic assembly of the recent past.

The integrated circuit shown in Figure 4.11 is fabricated on a minute semi-conductor plate, where the single components

Figure 4.11.
An integrated circuit fabricated on a minute semi-conductive plate

instead of being assembled manually are directly originated with diffusion processes in special furnaces. The connections between the various components are chemically engraved on the semi-conductor 'chip' itself, without any manual intervention, and thus provide a mechanical stability far greater than that provided by soldered connections. The connections between the active chip and the terminals of its supports are welded, and the assembly is hermetically encapsulated in ceramic material.

In addition to the drastic miniaturization, simplification is achieved because the designer has no longer to deal with single components (diodes, etc.) or the study of the elementary circuit but can devote himself to the conception of the equipment in its functional entirety. Thus the introduction of micro-logics is resulting in a more progressive step than that brought about by transistors replacing thermionic valves.

The micro-circuit illustrated contains no less than thirteen transistors, sixteen diodes and twenty-two resistors in a space of 3 mm $\times$ 6 mm $\times$ 18 mm, while the modular block, shown with the upper cover removed, replaces up to twenty printed-circuit cards of the conventional type.

DEA Block System

This has been designed in modular form to permit full exploitation of the potential data handling of micro-logic circuits. Figure 4.12 shows how the system is composed of standard

Figure 4.12.
*A modular design of
a DEA Block system
micro-circuit*

elementary blocks which can easily be removed from the cabinets for service or replacement and which can be complemented by others after a period of service. A particular feature is that the modular blocks interchange information amongst themselves through a coded transmission system. The blocks are interconnected in conference fashion through a few wires along which the data are exchanged in coded form.

Where a complex system is needed with the machine when it is new, a mini-computer may replace the standard DEA modules and may provide the same, or greater, scope of operations.

Mechanical Structure

The configuration of a measuring machine must change with the size of the workpiece, for as the machine becomes larger in capacity it is necessary to change to a free standing design in order to give the operator complete freedom of movement as he works around the component. On smaller models, a machine may include an integral surface plate made of either black granite or cast iron, while on the larger machines the decision as to the type of work support is left to the customer.

For efficient operation the following design elements are required.

(1) The machine must be independent of floor deformations.

(2) It must be independent of floor vibrations in both the vertical and the horizontal directions.

(3) Accuracy must be independent of the position of the sliding parts. This built-in vibration insulation eliminates the need for a special foundation which would otherwise be required.

Figure 4.13 shows the DEA patented design of structure incorporating the features mentioned; the accuracy obtained is independent of workpiece mass up to 3,000 kg, with operating speeds ranging from 10 μm/s to 5 m/min. To eliminate the risk of distorting the machine through external pressure, the machine motions are servo-motor driven and are operated from a movable-control pedestal.

Figure 4.13 indicates by heavy lines and shading how the machine is constructed. Figure 4.13(1) shows the three columns which rest on a concrete foundation, the machine stability being obtained by the three-point support. Figure 4.13(2) shows the main body of the machine including the table T which supports the workpiece but which is isolated from foundation vibrations since it is suspended from the three groups of indicated springs (these have a low resonance frequency). Figure 4.13(3) shows a plan view of the U-shaped structure which rests completely free on the main body

Figure 4.13.
*The construction of
a free-standing
measuring machine*

through three supporting points which are a spherical point, a linear roll guide and a ball bearing respectively. These points allow relative freedom of motion as indicated by the arrows. Therefore, when the workpiece load deforms the table, none of the deforming stress is transmitted to the slideways of the U-shaped structure, which is the real machine.

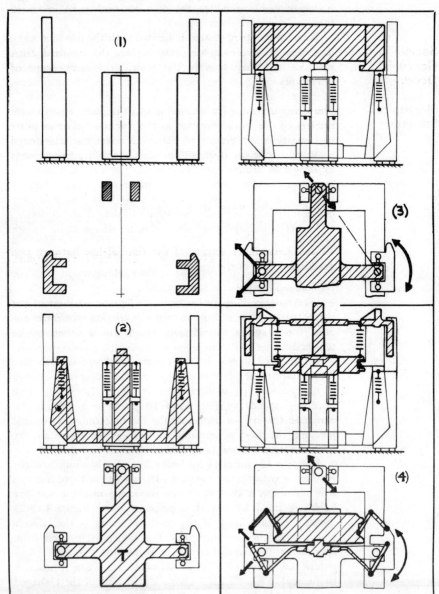

Figure 4.13(4) shows how the mass of the main and central carriages is completely supported by the pantograph system mounted on top of the two columns. In this way the main carriage is completely floating, free of the weight of the U-shaped structure, while the central carriage is similarly floating on the main carriage. Thus by the double suspension no sagging of X and Y slideways occurs, and therefore there is no error in work measurement.

Société Genevoise Developments

The firm Société Genevoise Ltd built their first measuring machine in 1909, and the present manufacturing programme covers practically all measuring and inspection requirements of modern industry. A typical example is the Trioptic universal measuring machine of Figure 4.14. Measurements can be carried out in linear co-ordinates in three axes or in polar co-ordinates with the use of a dividing table. The guaranteed accuracy of the machine without accessories is in metric units ($1\,\mu\text{m} = 0.001$ mm) as follows.

Reading
This term indicates the smallest sub-division of the micrometer screen and is 0.5 mm, while for estimation it is 0.1 mm. *Estimation* relates to the evaluation of the value corresponding to an intermediate setting of the fiducial line between two graduation lines of the fine reading dial.

Measuring accuracy
This varies according to the nature of the measurement, but the values indicated take into account the displacement dispersion error and the dispersion bias of the locating instrument which arises several times in the course of the same inspection.

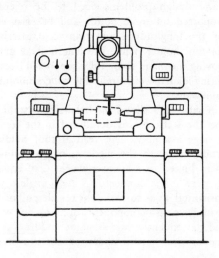

Figure 4.14.
A Trioptic universal measuring machine (By courtesy of Société Genevoise Ltd.)

Measuring accuracy with feeler microscope

Distance between mutually parallel faces	= 1.45 μm
Diameter of cylindrical bodies	= 1.50 μm
Hole centre distances	= 1.65 μm
Effective diameter of threads	= 2.15 μm
Vertical measurements	= 1.20 μm

Measuring accuracy with locating microscope

End standard measurements	= 1.50 μm
Transverse measurements	= 1.45 μm

Angular measuring accuracy

Circular dividing table, reading	= 1'' of arc
Dividing head, reading	= 10'' of arc
Goniometric microscope	= 1' of arc

The values given have been calculated according to gaussian principles and correspond to a probability of 95%. After the dimensions measured have been corrected by reference to the calibration chart of the standard scales concerned, the probability that the measuring error is smaller or equivalent to these values is 95%. The measuring accuracy can still be increased by computing the mean value of n measurements. It will then be practically equal to the relevant value listed divided by $n^{1/2}$ for a probability of 95%. It is necessary that the geometry of the component inspected is well defined, that its surface quality is good and that both standard scale and component temperature are in agreement.

Tool Holder Balancing System

To obtain high precision in building machines for inspection, many design problems need to be overcome. The bed is supported on three points and has flat and vee guideways for the longitudinal and transverse carriages, as well as for the tool holder, while the length of all guideways allows the moving members to be fully supported over their entire travel. Spring-loaded weight-relieving rollers minimize friction on the guideways.

Figure 4.15 shows the balancing system for the tool holder and the static forces developed in the transverse and vertical carriage assembly. The vertical tool holder saddle H is suspended through the medium of an inclined connector to a chain linked to a counter-weight C of equal weight, i.e. $P_1 = P_2 = -P'$. The connector is at an angle α with respect to the horizontal so as to make the tensile pull of the counter-weight chain develop a component force, $F_2 = P' \cot \alpha$. $F_2 - F_1 = F_1'$, and consequently the tool holder saddle is pressed continuously against the vertical slides of the transversal carriage, thus

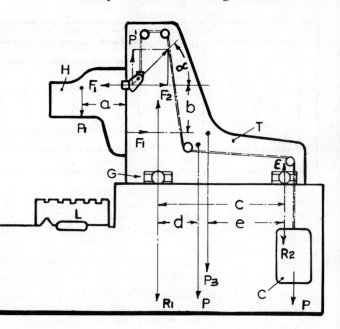

Figure 4.15.
*A balance system
and static forces of
measuring machine:*
C   counter-weight
G   releasing rollers
H   tool holder
     saddle
T   transversal
     carriage

ensuring displacements devoid of play. The four releasing
rollers G are so arranged that $R_1 - D = R_2 - E$, and the
friction effects on the slides are minimized and distributed
equally. The remote position of the counter-weight C at the
rear of the transversal carriage T results in a shift of the centre
of gravity of the assembly, thus providing better balance.

The moving members are driven by shafts, pinions and
racks; these are set so that their pull is exerted along an axis
close to that of the component of the friction forces.

**Positional
Readings**

Readings of the position of the three moving members are
made on built-in projection screens by merely centring the
image of a graduated line of the standard scale or of a fiducial
mark between two rectilinear and parallel lines of the reticule.
The optical reading system is free from parallax and functions
by dividing drums situated beside the screens that are mech-
ically driven by the longitudinal and transverse carriages. They
are graduated in millimetres, and the reticules of the screens
carry eleven pairs of setting lines and are adjusted by knobs
in order to 'frame' the scale projected line. Since there is only
one line to be viewed at a time, any reading ambiguity is
obviated. Readings to 0.1 mm are made on the reticule
graduations, and the displacement of the reticule imparts
a rotation of the circular dial carrying 200 divisions reading

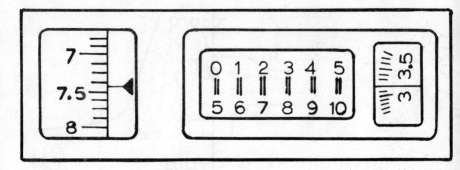

Figure 4.16.
*A projection screen
for the three moving
members of the
Société Genevoise
machine*

each to 0.0005 mm. Figure 4.16 shows the projection screen, the reading example being 173.6325 mm which is made up of 173.0 for the drum, 0.6 for the reticule and 0.0325 for the dial.

The best inspection machines are limited in use without a comprehensive set of attachments able to solve the inspection problems. One of these is the feeler microscope shown in Figure 4.17(a), (b). The displacements of the feeler are optically magnified and viewed on a projection screen. A fiducial mark engraved on the screen provides an accurate reference position of the feeler. The microscope lends itself

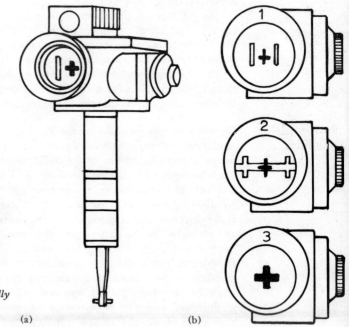

Figure 4.17.
*The feeler
microscope optically
magnified for
projection*   (a)   (b)

to the following modes of operation.

(1) Figure 4.17(b) shows it used with the measuring pressure giving lateral movement without vertical displacement of the feeler, e.g. to measure distances between mutually parallel faces, diameters of cylindrical bodies, hole centre distances, hole position in relation to a drum face, internal and external tapers, testing alignments.

(2) It can be used with measuring pressure giving vertical movement but without a vertical measuring pressure, e.g. to measure the pitch and diameter of cylindrical or conical, internal or external threads, the conicity of tapered internal or external threads and their diameter in relation to a face.

(3) It can be used with a vertical measuring pressure and with vertical and lateral displacement of the feeler but without lateral measuring pressure, e.g. to measure the pitch of racks or of external threads on parts held between centres or on V blocks.

The tube of the microscope can swivel through 90° to cater for measurements in the X or Y axis, while the screen of the microscope remains facing the operator. All feelers are fitted with an automatic over-ride to prevent damage to the microscope through accidental mishandling.

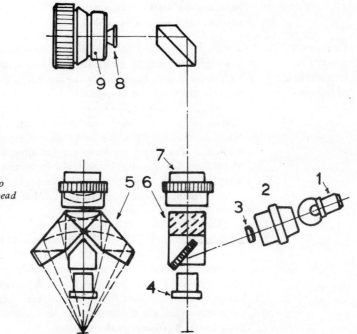

Figure 4.18.
An attachment to the micrometer head for measuring without contact:

1 *source*
2 *condenser*
3 *signal*
4 *objective*
5 & 6 *prisms*
7 *objective*
8 *adjustable reticule*
9 *eyepiece*

Measuring Without Contact

Irregular surfaces such as turbine blades, aerofoils or die cannot be measured by means of the feeler microscope because of the undefined position of the contact point between the feeler and the surface under inspection; nor can the feeler microscope be used for the measurement of soft parts such as wax models or plastic components because small as the measuring pressure is, it can cause deformations that result in measuring errors. The problem is solved by using an attachment mounted on the micrometer head, its functional principle being indicated in Figure 4.18.

The light beam emitted by the source 1 and passing across the condenser 2 projects the image of a cross (signal 3) onto the objective 4 and then onto the part to be inspected. The image of the cross is reflected through the prisms 5 and 6 and the objective 7 onto the eyepiece 9 with the adjustable reticule 8. If the distance between the microscope and the part is correct (accurate focusing), a single image appears on the eyepiece. If the distance is too short or too long, there are two images. The successive vertical positions of the microscope are read on the corresponding screen of the machine.

The dispersion that affects measurements is small when carried out in this way, even if made on surfaces inclined up to almost 70° from the horizontal. It depends on the surface quality of the components to be measured but generally oscillates between ± 1 μm to ± 3 μm. Microscope magnifications are 24X or 48X.

Converting Jig Borers to Inspection Machines

Sogenique (Services) Ltd specialize in rebuilding SIP jig borers, in converting them to measuring machines with accuracies of a high order and in maintaining the quality which has made the original machines world famous. A wide range of equipment can be fitted such as digital read-out display, printer recording electronic 'XY' axis error computing centre equipment and other accessories from the range of SIP standard measuring machines. The basic machine can be supplied in a number of versions from a simple two-axis drum and vernier-reading machine using master screws and a simple revolving measuring spindle to the sophisticated three-axis machine with a motorized travelling table and cross beam.

Figure 4.19 shows a typical conversion to one of the larger machines; this has a longitudinal table travel of 1,300 mm and a cross travel of 1,000 mm (*X* and *Y* axes respectively). All machines are rigid; the one shown weighs 5,400 kg and is equipped with a transducer system which makes it independent of the measuring screws, but even when traverses are dependent on screw operation the master screws used give a measuring accuracy of 0.007 mm peak-to-peak.

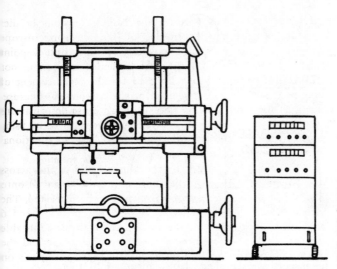

Figure 4.19.)
*A jig-boring machine
converted to a
measuring machine
(By courtesy of
Soginique Ltd.)*

The large capacity of these machines and the original quality makes these converted machines an attractive and economic proposition.

**Gear Tooth
Measurement**

In view of the vast quantities of gears produced and the increased accuracy demanded owing to high-speed operation, gear inspection assumes a primary place in engineering. For general checking purposes of the chordal thickness, the gear tooth caliper is available, but the use of the tangent comparator has increased because measurements can be made without reference to tip circumference. It will be appreciated, however, that hand methods of checking are not suitable where large quantities of gears are manufactured, and gear-testing machines for checking all types of gears are available for rapid inspection of tooth profile, eccentricity and centre distance. Again, where very fine limits of accuracy are required, optical and electronic means are finding application in gear tooth measurement.

Goulder Mikron Ltd, Kirkheaton, Huddersfield, have patented new measurement techniques to permit improvements in the accuracy of gear measurement, allowing gear parameters to be measured in units of $0.25\,\mu$ft. One of the arrangements enables two or more motions (or other parameters which can be expressed as electrical frequencies or pulse trains) to be compared accurately, and it is considered that this technique could have applications in fields other than gear measurement.

Involute gears, and more particularly the involute helicoid gears, are of complex three-dimensional forms which have to be measured to fine limits. The surface shape is defined by an equation that is linear — though complex — and mechanisms

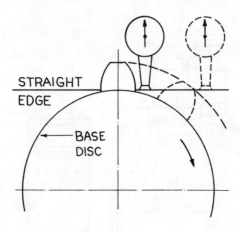

STRAIGHT
EDGE

BASE
DISC

Figure 4.20.
*The operating
principle of an
involute gear tester
(By courtesy of
Goulder Mikron Ltd)*

can be constructed both to generate the required profile and to inspect the finished shape. Figure 4.20 represents the operating principle of a simple involute test in which the gear is mounted on a common axis with a disc of diameter equal to the base circle diameter of the gear. A straight edge in contact with the disc carries a measuring stylus which by deflection can be used to show the error on a dial indicator.

This system suffers from possible sources of error such as inaccuracies in the base circle disc and straight edge with the possibility of slip in the friction drive, deflections caused by the use of contact pressures sufficient to avoid slip and the need for a different base circle disc for each gear size. Although more complex systems based on the same principle can be designed to minimize the disadvantages, the measuring accuracy remains limited by mechanical considerations. In an alternative mechanism designed by Goulder Mikron, the stylus is replaced by a rigid 'pusher', and when the slide is moved tangentially to the base circle this pusher engages with the involute profile to rotate the gear. No base circle disc is required as the drive is direct to the gear. According to the definition of the involute form, if the straight edge is moved at a constant linear velocity, the gear will be rotated at a constant angular velocity.

There is thus a fixed relationship between the linear velocity of the slide and the angular velocity of the wheel, and by measurement and comparison of the instantaneous values of these velocities any departure of the tooth form from the true involute can be detected. A method of measuring the deviation was developed, using optical gratings. As shown in Figure 4.21 a linear grating A is used for the slide motion and a rotary grating B for the gear movement, and it is necessary to

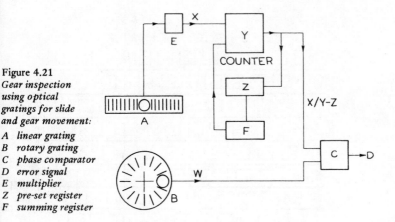

Figure 4.21
*Gear inspection
using optical
gratings for slide
and gear movement:*

A   *linear grating*
B   *rotary grating*
C   *phase comparator*
D   *error signal*
E   *multiplier*
Z   *pre-set register*
F   *summing register*

compare the pulse frequencies obtained from these grating systems to an accuracy of the order of one part in a million. Since the frequencies x and w are not, in general, related by a simple integer multiple, a special digital-processing system had to be devised.

Pulses from one grating are fed directly to a phase comparator C, and the train of pulses from the second grating is processed to convert it to the same frequency w as the first signal; this is fed to the phase comparator. The latter derives the phase difference between the two inputs as an error signal D which is supplied to a chart recorder. The frequency conversion ratio depends upon the gear dimensions and grating characteristics and is a non-integer which can be defined to seven or more significant figures. Typically, the grating may have 1,000 lines/25 mm, and the transmission ratio may be 1:0.3145138; the pulses are fed to a ×1,000 multiplier E to enable changes of one part in a million to be detected. In this case the frequency ratio will then be 314.5138, and this ratio may be expressed as yz, where y is the integer portion 314 and z the decimal part 0.5138. Z and F represent the pre-set and summing registers.

Optical gratings are also employed in a technique developed for the measurement of gear pitch errors. In this system, the gear is rotated continuously, and a stylus is inserted and withdrawn between the teeth with the point of measurement being defined by reference to the grating which rotates with the gear. The stylus is introduced just before the measuring point and is deflected by the gear tooth through the measuring position before being withdrawn. The output from the transducer actuated by the stylus is of a ramp form (Figure 4.22), and measurements of the output are triggered by the grating

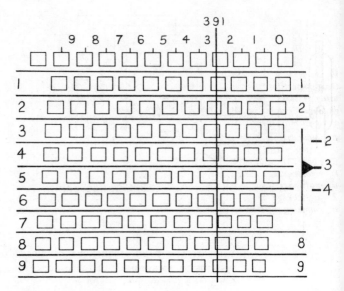

Figure 4.22.
*The ramp form of
the output
obtained from
a transducer*

system at intervals corresponding to the gear pitch angle. Pitch errors are indicated by the difference between the values obtained for successive teeth and can provide a record of adjacent and cumulative errors.

Machines will accommodate gears with diameters from 25 to 1,016 mm and can be used for checking gears varying from 5 to 999 teeth. Angular accuracy is ± 1" of arc, with the speed of testing 2s/tooth.

Radiographic Inspection Equipment

With this system the detection of faults is effected by comparing the electrical signals transmitted by a television camera, with standard reference values, and the system described is a development of Usines Balteau of Belgium. Figure 4.23 shows a disc A to be inspected; it is loaded onto two supporting rollers B which cause it to rotate, and this mechanism is fitted between the X-ray source C and an image-intensifying amplifier D. The X-rays pass through the disc and a radioscopic image is formed on the screen K of the image-intensifying amplifier. The screen, with other components, is enclosed in an evacuated glass envelope. It is charged to a negative potential of 30.000 V. The second screen L is at zero potential.

The screen K is composed of a layer of material which fluoresces under X-ray radiation, and a layer of photo-sensitive material which emits electrons in proportion to the degree of illumination. These electrons are accelerated by the potential difference and are electrostatically focused

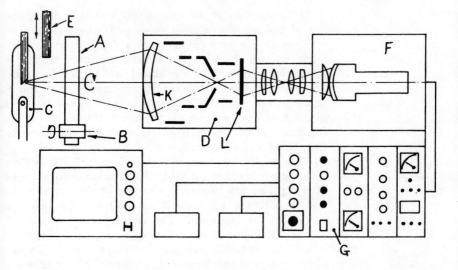

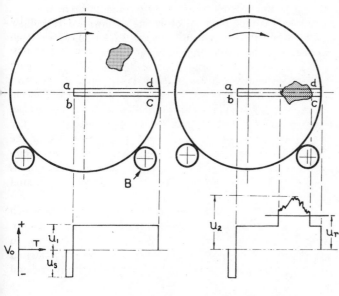

Figure 4.23.
Fault detection by electric signals from a television camera

and directed onto the screen L to reconstitute an image which is 6,000 times as bright as that initially formed on screen K. The television camera F converts this image into electrical signals which are transmitted to the control unit; this is adjusted so that it analyses the area limited by the window abcd in Figure 4.24, and one revolution of the disc serves for its inspection.

Figure 4.24
Inspection of a disc during one revolution using automatic operation:

A   *disc (Fig. 4.23)*
B   *supporting rollers*
C   *X-ray source*
D   *image-intensifying amplifier*
E   *lead shutter*
F   *television camera*
G   *control unit*
H   *monitor with screen*
K   *screen*
L   *second screen*

If the variation of the video signal with time during one line of horizontal scanning is considered, a zone that has no fault, as at the left, will result in a negative value, which is the level of the horizontal synchronization. No image is formed because an image is produced only by positive values. This negative value is followed by a positive value u_1 corresponding to the level of illumination of the screen with respect to the zone analysed. For the position of the disc at the left the zone exhibits no fault and is of uniform thickness, so that the value of u_1 is constant.

When the disc is in position at the right, a fault is coincident with the window, and the value of u increases when the spot of the cathode-ray tube scans the zone of the image of the fault, because the latter absorbs less radiation than the sound portion of the component. This increase results in a higher degree of illumination of the fluorescent screen of the image-intensifying amplifier. The positive voltage now reaches the value of u_2, and this increase is compared with the reference value u_r and can be used to set in operation a mechanical arrangement for sorting the satisfactory components from the faulty ones.

The value of u_r is determined by observing a selected area of the image that corresponds to a constant absorption of radiation. This is to compensate for variations in the elements which contribute to the image and which are not connected with the fault. During the transfer of the discs, the lead shutter E takes up a position between the X-ray source and the image-intensifying amplifier, in order to avoid subjecting the latter to an excessive intensity of radiation. The television monitor H is used only for observing the image during adjustment. When the equipment is in use, such observation is not necessary as the operation is fully automatic.

5.
Probes for Measurement & Inspection

Introduction

To take full advantage of the remarkable accuracy of modern inspection machines, a range of universal or custom-designed tooling equipment is required. This comprises the availability of a range of various types of probe along with certain accessories for calibration. Figure 5.1 represents a typical set of probes with additional equipment complementary to their use. Figure 5.1(a) shows a universal swivelling dial indicator probe, while Figure 5.1(b) shows a granite square for calibration. Figure 5.1(c) is a reference gauge for tool qualification and Figure 5.1(d) a fixed probe. For checking the angles of special probes the calibration instrument of Figure 5.1(e) is available, while Figure 5.1(f) is a spring-loaded scribing tool used for marking-out purposes on irregular surfaces. Figure 5.1(g) shows an articulated scriber with Figure 5.1(h) indicating a template scanning device. A projection microscope is featured in Figure 5.1(i) and Figure 5.1(j) indicates two types of tapered probes with a larger tapered probe in Figure 5.1(k). Two types of balltip probes are shown in Figure 5.1(l).

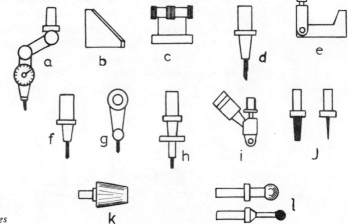

Figure 5.1.
*A typical set of
probes used on
measuring machines*

Use of Probes

The probes shown in Figures 5.1(a) to (l), by no means show the whole range of probes used, and many others are described later, but the simplest measuring probe can be considered as a short length of rod with a hemisphere on its end placed vertically in the ram of a measuring machine and brought into contact with the workpiece. At the first point, the read-out will be made to read zero, and when the probe is moved to the second point the read-out will show the different position taken up by the X and Y axes, together with the difference in height between the two points. The information is presented in this way because it is necessary to 'position' on two axes with the measuring machine and to measure with the third axis as shown in Figure 5.2(a).

With equal facility a conical probe can be fastened vertically in the ram and in turn inserted in each of two holes in a component. The read-out displays give the distance between the two holes on X and Y axes if the read-outs were 'zeroed' at the first hole. The 'Z'-axis reading in this case is meaningless as it only serves to show how deep the probe has entered the hole (Figure 5.2(b)).

Both these systems of probing have severe limitations: in the first case, apart from the normal inaccuracies in the machine, the amount of pressure applied to the probe will have some effect on the read-out figure because of deflections in the machine structure. A soft workpiece may also deflect under probe pressure. On all but the smallest of machines it is difficult to provide both a stiff machine to withstand the load and to give linearity of movement and a machine which is sensitive enough to move under the slightest pressure from the probe with attendant sensitivity of touch.

Equally the conical probe can introduce further errors, a coarse feed scroll in the bore of the hole can make the hole entry 'out-of-round'. A chamfer on the hole may not be co-axial with the hole, and burrs thrown up or even material deformation in drilling can all contribute to introduce errors; finally, if the hole is not perpendicular to the upper face, the read-out position will not be correct. This is not to suggest

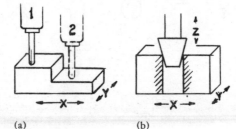

(a)   (b)

Figure 5.2.
*Probes measuring
(a) flat surfaces and
(b) a conical bore*

that simple probes should never be used, for there are numerous applications for their use, but where high precision on large machines is required more elaborate probes are required. Thus for contour checking a simple fiducial or dial indicator will prove capable of eliminating errors of touch pressure. A proximity device can also be used to eliminate any contact between probe and workpiece. An instrument of this sort which measures the electrical capacity between probe and workpiece was developed by Doncaster-Monkbridge Ltd. of Leeds and has been used continually with a variety of probes to measure complicated turbine blade dies. With a near-spherical end to the probe it becomes a three-directional instrument.

DEA ELECTRONIC FINGER

Figure 5.3.
A DEA electronic touch finger TF2 for high-speed measurement of shoulders during movement

DEA have developed an electronic touch finger which is shown in Figure 5.3. Basically this consists of a fixed and a sliding member in which the latter carries a pair of inductive elements. The sliding member is brought into contact with the work and is progressively advanced so

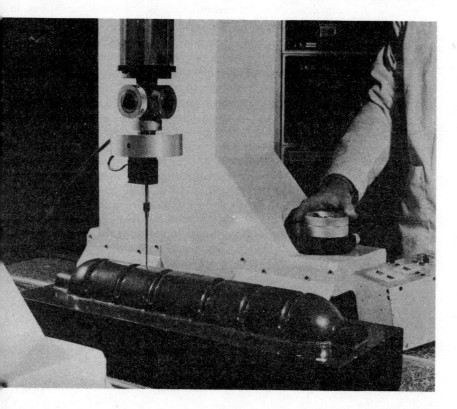

that it slides within the fixed member. At a precise point in its travel a bridge circuit is brought into balance, and a memory unit in the control console holds the figures displayed in the read-outs. They are then automatically recorded either by electric typewriter or by punched tape, or by both.

The essential feature of the touch finger is that it measures whilst it is moving, thus saving time on each measurement by eliminating slow positioning of a fiducial or other indicator; it records the co-ordinates of the point measured, whilst the operator is moving the machine to the next point to be measured. The operator knows that a measurement has been made as an audio-tone is emitted by the control equipment simultaneously with the measurement taking place. This tone remains as long as the sliding member is beyond the point of balance. The probe may face in any one of five directions when mounted in the machine; it can therefore be used for certain measurements in X and Y axes as well.

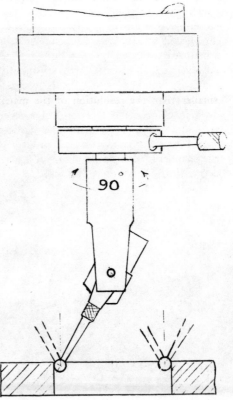

Figure 5.4.
An electronic finger for hole location and for shoulder and groove measurement

PRECISION HOLE CENTRE LOCATION

As a further extension of the same principle, DEA also make an electronic touch finger for precision hole centre location. Its uses also include co-axiality checks and shoulder and groove measurements. Figure 5.4 shows the probe with its articulated finger which carries the inductive element that in this case senses the point at which the touch finger is precisely parallel with the fixed axis of the probe body. The complete probe is made to index through 180°, and thus, by touching the finger against one side of a hole, by swinging the finger through the balance point and by repeating this at the other side of the hole, two points of a chord of the hole are retained in the memory unit on, say, the X axis. An algebraic unit subtracts the smaller from the larger, divides the difference by two and adds it to the smaller figure, thereby printing out the ordinate of the hole in the X axis. The probe is then indexed through 180°, and the process repeated to give the Y-axis ordinate. The complete process takes only 2 to 3 s and is shown diagrammatically in Figure 5.5.

A portable control board is provided with the electronic touch finger, and by means of selector switches it is possible to select other modes of operation such as checking the overall dimensions of a shoulder or the whole component or measuring a staircase arrangement or hole diameters. In the latter case, however, a high order of accuracy is difficult to obtain owing to the possibility of the sum of several small errors, each smaller than the resolution of the machine. For instance, if on a machine that has a resolution of 0.05 mm an error of 0.04 mm occurred in finding one of the ordinates, then there would be two further errors of up to 0.04 mm each in the diameter due to probing at two points not exactly diametrically opposite. This kind of error becomes worse as the hole diameter becomes smaller, and coupled with it

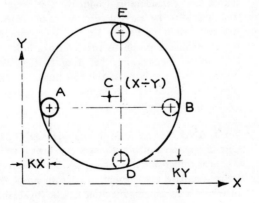

Figure 5.5.
*The process of
probe measurement
using an electronic
finger*
$CX = B\text{-}A/2 = K_X$
$CY = E\text{-}D/2 = K_Y$

there are two further possible errors of up to 0.04 mm at the points of probing. However, on very-small-diameter holes with close tolerances there are many other convenient ways of checking.

On the other hand, the electronic hole centre probe offers several distinct advantages over the other systems in that one probe can cater for every size of hole from, say, 2 mm in diameter up to the maximum capacity of the machine; it can also inspect hole centre distances deep down in holes or recesses, thereby also checking squareness to the datum face. Recesses, multi-diameter holes and conical holes can all be checked by fitting a ring to the tip of the probe instead of the more normal spherical end.

A special probe has been in use for some time by the L.K. Tool Co. Ltd. which is designed particularly for use in conjunction with a mini-computer. It can be used for surface, edge-to-edge or hole measurement, and this versatility frequently permits complicated components to be checked completely without a probe change. The probe is spherical and can move axially or in any other direction according to the pressure applied. Hole position and diameter measurements can be made by touching the bore at three points. The computer rapidly calculates the one diameter which will fit the three points together with the hole centre position. The PDPII and Hewlett-Packard mini-computers are extremely adaptable to use with this probe.

Range Extenders

Ferranti manufacture a series of conical range extenders as shown in Figure 5.6(a). Three rollers are held in the bore to be measured, while a smaller conical probe is inserted in the centre, causing the rollers to move out to contact the side of the bore, thus centralizing the probe in the middle of the bore. The attachment is particularly useful when measuring large diameters, considerably reducing the weight of the probe which has to be carried by the machine. A similar Ferranti device, shown in Figure 5.6(b), is comprised of ball bushes which enable a probe to reach into a hole away from burrs, bell-mouthed edges or other irregularities. When the probe is inserted into the bore of the ball bush, the three captive balls move out to contact the side of the hole, thus centralizing the probe in the middle of the hole.

TWO- AND THREE-DIMENSIONAL CONTOUR CHECKS

One of the most difficult inspection jobs is to be found in two- or three-dimensional contour checks. For the former, a copy lathe template can present inspection problems and is an area in which a measuring machine can be invaluable. In most cases

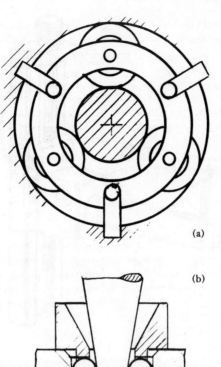

(a)

(b)

Figure 5.6.
*Ferranti conical
range extenders
using ball bearings*

Figure 5.7.
*A probe with a
knife edge which
measures by contact
with the template*

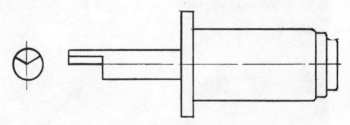

a probe as in Figure 5.7 is used. It can be seen that the knife edge is used for measurement by contact with the template and that the support for the knife edge may be rotated to clear parts of the template while still leaving the knife edge on the centre-line of the measuring machine ram.

Cams, both flat and cylindrical, may be inspected in this way, and an alternative design is available to perform either two- or three-dimensional contour inspection (Figure 5.8).

Figure 5.8.
*A probe with sharp
point and vernier
setting*

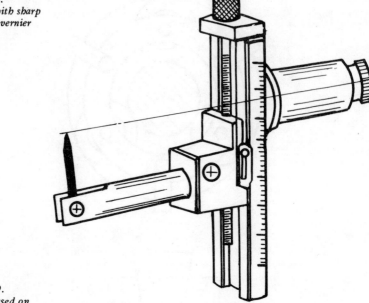

Figure 5.9.
*A probe used on
motorized machines
with control from a
punched tape*

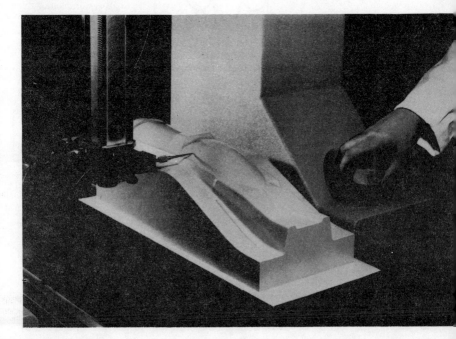

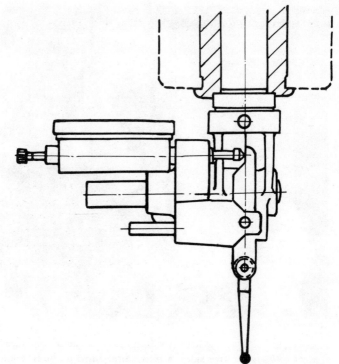

Figure 5.10.
*A low-cost probe
with a dial indicator
for checking bores*

The probe employs a sharp point which can be set on the centre-line of the machine ram, and its support can be rotated to allow access to the component from any angle. The probe can accept a dial indicator instead of the sharp point which gives accurate end results but which may take a little longer to operate.

For surface or contour measurement the probe shown in Figure 5.9 is ideal. It is often employed on motorized machines with control from a punched tape to scan a large-contoured component, taking checks at pre-determined points and recording the information on either another tape or by print-out. The machine operator has little to do except to watch the operations from a desk via a monitor or to stand by the probe.

Where low-cost mechanical hole measurement is required, a device using a dial indicator and a probe can give excellent results. A typical probe is shown in Figure 5.10, the probe being rotated and the position of the machine ram adjusted until a 'zero' reading is obtained. The machine will then show the co-ordinates of the hole relative to a previously fixed datum and will also show the 'out-of-roundness' of the hole.

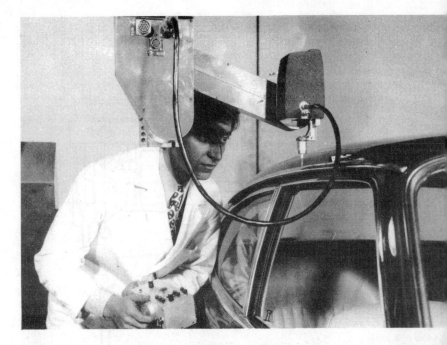

Figure 5.11.
A probe with triple articulation used in conjunction with a computer

Where it is important to check contoured components from five sides, a multi-directional probe is required, but some form of directional adjustment may be needed during the measurement of each side together with an angular approach on occasions. Figure 5.11 shows a probe which can fulfil these requirements and which is used on a very large DEA machine intended to measure car body models. The probe has triple articulation and must be used in conjunction with the computor which automatically compensates for any movement of the articulated joints by a corresponding alteration in the readout/print-out figures to show the true position of the tool point at any time. These movements are motorized and can be manually made or controlled from a tape during automatic scanning. The actual measuring head uses the familiar differential transformer-type transducer.

Certain small features on components such as tiny holes or scribed lines can be measured very successfully using a projection microscope. With suitable magnification, both the position and the diameter of small holes can be recorded easily, and an estimate can be made of any out-of-roundness. This can also be a good method of setting to a machined edge to use it as a datum. The various features of the component can be clearly seen on a screen which has both polar and cartesian graticules for ease of measurement.

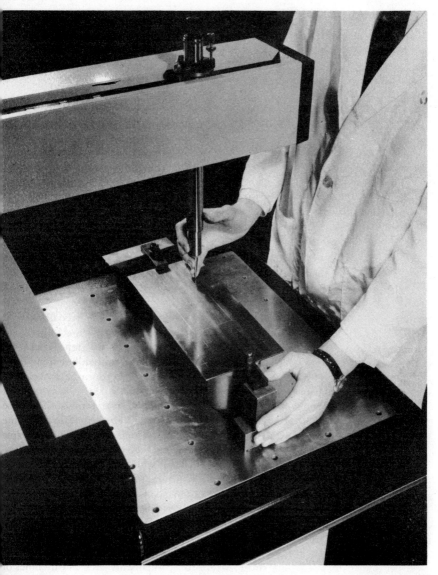

Probes for Marking Out

Figure 5.12. above
An example of a probe being used for marking out

Essentially, marking out consists of scribing a series of lines or circles on a casting or similar part as a guide to the machinist while setting and cutting the component. It can also be seen as an inspection operation to prove the quality of goods arriving at the factory. The probes consist of a series of sharp-pointed instruments which can be mounted in the ram of the machine and which can be moved over the component to produce the lines as indicated in Figure 5.12.

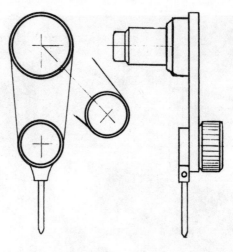

Figure 5.13.
*A scriber point
carried on an
articulated arm*

The simplest tool consists of a pointed plunger which is spring loaded and operated in the manner shown. However, where castings are rough or where the surface to be marked is at an angle to the axes of the machine, a different approach must be made. An articulated probe is available to enable both rough and angular faces to be marked out without difficulty. It consists of a scriber point carried on an articulated arm (Figure 5.13). The ram of the machine is positioned, and the scriber arm is moved by hand, but little or no load is placed on the frame of the machine using this tool.

For marking circles a typical universal tool is shown in Figure 5.14; this is provided with a calibrated scale to set to the diameter of the hole to be marked out, and the position of the hole is set on the machine read-out. It can be seen that this marks the periphery of holes which lie on faces at an angle to the axes of the machine. Again, it is somtimes necessary to make a more permanent mark than a scribed line for the hole centre, and this can be achieved by mounting an electric or pneumatic drilling head on the ram of the measuring machine. A small centre drill can then be used to provide a guide for the drill operator.

Turbine and Compressor Blade Measurement

Extremely accurate profiles are required in the gas turbine industry where blades are concerned, and numerous systems have been evolved to ensure the accuracy of blade dies, for forging, and of any subsequent machining of the blades themselves.

A method used for many years, for both blades and dies, has been to make precision templates which are mirror images of the profile at particular points along the blade or die. The

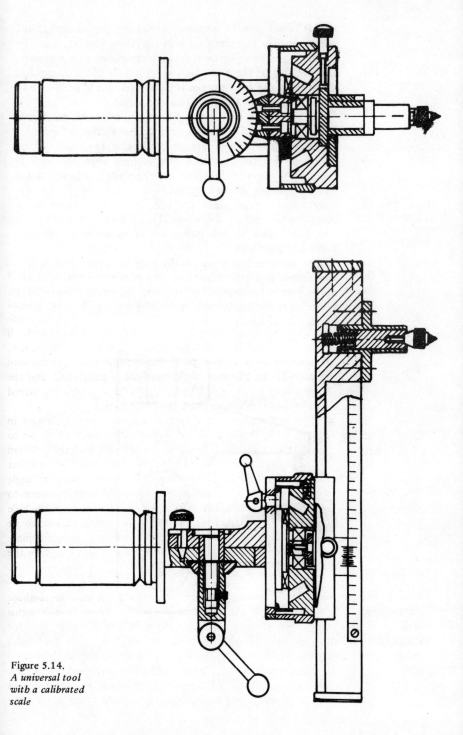

Figure 5.14.
*A universal tool
with a calibrated
scale*

profile could thus be inspected to see if a gap existed at any point between the template and the blade profile. If a high-precision profile was required, many templates were necessary, and the cost of these was high. Additionally the inspector had to estimate the gap between template and blade or die, and a direct reading was always difficult to obtain.

Machines have been pressed more and more into service for blade inspection, and probes such as that shown in Figure 5.3 have been used for this purpose. Whilst this represented a large saving in templates, at least two settings were necessary to check all surfaces of the blade, and the blade had to be rotated by some means through approximately 180° to examine the convex and concave sides. The L.K. Tool Co. Ltd have developed a probe which eliminates the need to rotate the blade and which permits fast inspection of as many points as is desired (Figure 5.15).

The upper half is a known distance from the lower half of the probe, and after inspecting the convex surface with the upper part of the probe the computer-calculator automatically subtracts that distance from its reading when making thickness checks on the blade.

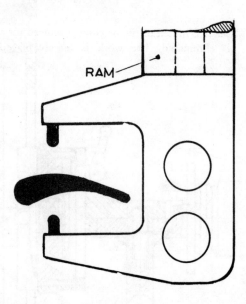

RAM

Figure 5.15.
A probe for turbine blade measurement

6.
Computer–Aided Inspection

Introduction

References have already been made to computerized inspection on Ferranti and Olivetti machines, but this chapter extends the information to inspection on large and complex components mainly in the automobile industry where it is now being realized that the inspection machine can be of real value. However, to commence with a simple example and to build up from that, one of the most time-consuming operations in arranging a conventional inspection set-up is aligning the workpiece with the datum axis of the measuring equipment; even when the machine table can be adjusted to trim the alignment, the time can be lengthy. When, however, a computer is linked to the machine, the need for accurate alignment is eliminated. The work is merely placed on the table in

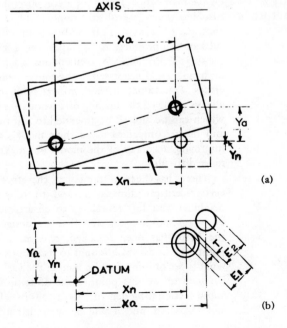

Figure 6.1.
Two datum holes being probed by means of a computer

approximately the correct attitude, and by specified datum points being probed the information can be generated from which the computer automatically calculates the amount of misalignment and makes compensation to all measurements subsequently obtained during the sequence.

Figure 6.1(a) shows how the two datum holes may be probed when the work is in position on the table. If alignment were perfect, the X co-ordinate X_n would equal the centre distance, and the Y co-ordinate Y_n would be zero, but misalignment is detected because the former is less than the centre distance, shown by X_a, and the latter has a finite Y_a. The computer calculates the misalignment angle and applies it to all measurements made subsequently, so that in effect the work is 'viewed' correctly by the computer, in the direction of the arrow. Figure 6.1(b) shows another computer manipulation whereby a hole is detected that is misplaced, falling on the co-ordinates X_a and Y_a instead of the correct co-ordinates X_n and Y_n. The computer will work out the polar error E_1 or, if required, the calculation can also include the amount by which the hole is outside the permitted tolerance T, as indicated by E_2. A further facility is that of probing three points spaced at random around the periphery of a circular form, in order to obtain data from which the computer calculates the radius, diameter or position of the centre.

Car Body Manufacture

Of the many applications for a computer linked to a measuring machine there is probably none more complex than the inspection of a motor car body or aircraft models with the subsequent production of tapes for a continuous-path NC die-sinking machine. A tremendous quantity of data has to be handled and processed in the shortest possible time without errors. Traditional methods of car body manufacturers have always involved the making of a wood, plaster or plastic model which can be altered or amended to suit the designer's wishes, and many companies make their models to a 1:1 scale, thus improving accuracy of the model but making its measurement more difficult.

There has been a tendency therefore only to measure certain critical dimensions and to rely upon copy-milling machines and loft templates to ensure that the finished car body resembles the shape that its designer intended. After the shape has been finalized, a decision is taken at which points the body shall be split to its various separate panels and what types of joint shall be used at the various panel junctions. Frequently, at this point, a set of templates is made which allow sectional models of the various panels to be constructed and inspected, and it will be seen that up to this point very little dimensional information is needed. Usually, a copy-milling machine with a 'reverse' image facility is used to make a female die from each section model, again without a significant amount of dimensional information.

Normally, another die or punch is required as well, and often a further section model is made with allowances for the thickness of the sheet metal which is then turned into a male punch by a copy-milling process on a block of die steel. With the complete die set it is now possible to run a press test to produce a short run of prototype body panels which again are not normally measured but which are brought up against one another to test the matching of the joints and which are compared with the original loft templates. Modifications are usually necessary to correct clearances for the sheet metal and to give a better metal flow in the die set; not infrequently alterations are needed to bring the joint faces together to make a better joint, again without much measurement. There is little wonder therefore that one well-known British luxury car has been manufactured for several years with one side of its body approximately ¾ in. longer than the other side.

Functioning of the Body

The body has so far only been considered as a shell, but it must also function as an anchor point for all the other parts of the car. The suspension and steering parts must all be attached with precision to give good road-holding; and the engine and transmission must also be located properly to give long bearing life with quiet running. The motor industry has long considered that manufacture of the body and its anchor points for the other parts has been a more difficult task than many others associated with mechanical parts.

Indeed, two to three years is a typical time span between approval of the model shape and actual preparation to make bodies in production quantities. It will be obvious that many man hours are needed to check the dies and finished body and to see that all the other parts fit easily. Fascia panels and interior trims must fit well the first time that they are offered up to the body; a motor car assembly line is no place to start filing a corner from a door panel. Thus, if doors, boot and bonnet lid and other loose panels could fit immediately without adjustment (e.g. elongated holes), benefits would accrue on the assembly line from such precision.

It was against this background that two Italian engineers left the Fiat body plant in 1960 to form the company known as DEA of Turin to manufacture computer-controlled measuring machines which would speed up the process of car body manufacture and which would provide a degree of precision hitherto

Figure 6.2.
*A computer-
controlled
measuring machine
used in car body
manufacture*

unknown. The first machine known as the 'Alpha' has progressed through various stages until it now appears as shown in Figure 6.2; the details of the design have been given before. Servo-motors have been provided for all machine movements which has enabled tape controlled inspection of the dies to be undertaken. Additionally the continuous path milling of the models and probes has been developed to remain in contact with the part at all times during continuous-path scanning and these probes control the input signal for the servo-system. DEA had in fact created their own computer, the DEAC 1001, to process all the information necessary as it was not possible to buy a computer which would perform even a few of the functions needed by the Alpha.

The computer is a stored programme device which is capable of accepting a new programme card at any time in the future. It has a high-capacity memory with tape read, write and search facilities. Normally, in the inspection mode it records three-dimensional co-ordinate points at either predetermined distances in a matrix or at pre-determined changes of dimensions on one or two axes. To consider its functions best, a study of the processes in the manufacture of a car body will be considered using the DEA Alpha instead of manual skills.

When the mock-up model for the body has been made and finished, it is taken to the machine and is sited on the surface plate. Mis-alignment of the model can be accommodated on more than one axis, and the computer automatically compensates for misalignment. An electronic probe is inserted in the head of the machine, and a programme is selected which instructs the machine to scan the body at fixed increments on one axis; it travels on the other axis and measures the third axis but records all co-ordinates. A short instruction tape is inserted in the tape reader which defines the areas which are to be examined, whether it is the whole body or only the roof. Various programmes are available; the simplest checks at points 10 mm apart on the X and Y axes, and another records points closer together when there is a rapid change in dimensions on the Z axis such as at the corner of a roof. It is clear that, where contours change gradually, rather less information needs to be collected than that from areas where there is a sudden change in profile. Often, from the middle of the roof, for instance, points are only recorded at 25 mm intervals on the X and Y axes. While scanning is taking place, the operator can intervene at any time and perhaps pay particular attention to some point which his experience tells him may give trouble in the die manufacture.

While scanning is proceeding, the tape punch, driven by computer, is recording all the relevant information, and, after scanning is completed, a tape is immediately available for further processing. For instance, aerodynamic experts may wish to study the shape in detail, and a programme is available to extract information from the master tape which can drive a drawing machine (Figure 6.3) and which will draw sections through the car body in any one of three planes, while a dimensional print-out can also be obtained. Further post-processing can be undertaken; the car body can be divided up by its joint lines, and a sub-tape can be produced for each of the smaller panels which make the whole body. Mirror image tapes can be produced just as quickly, and feed-speed information for the NC die sinker can be inserted. The computer is able to derive dimensional information for points in between those at which it has taken measurements by both parabolic and cubic interpolation, a facility not readily available from any other source.

At this point in the sequence of events it is normal for the die-sinking machine to start work roughing out the die blocks whilst final consideration is being given to, say, the aero-dynamics. If it is considered that changes in shape need to

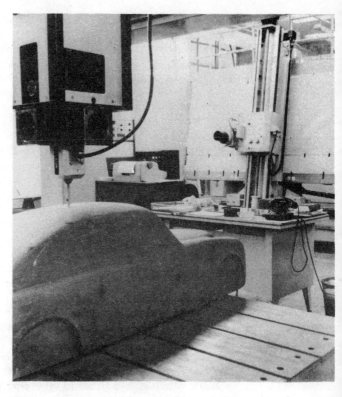

Figure 6.3.
*A machine used for
drawing sections
during car
inspection*

be made, the information can be given to the computer in the form of a short tape, and a milling head can be brought into use on the Alpha ram to modify the model. The computer will quickly re-process such tapes as need amendment, and the process of die sinking can continue to its logical conclusions. The computer is capable of proportional working and can therefore accept information from scale models and scaling up or down as may be required.

**Automatic
Scanning**

On completion of the machining operations, the die blocks can be scanned automatically by the measuring machine and its continuous-path probe, the computer making a direct comparison with the original tape. The scanning programme in this case can contain a tolerance which applies to all measured points. The print-out is usually of the form of theoretical dimensions shown against actual dimensions, with the difference between the two calculated and also printed. Any point which is out-of-tolerance is printed in red as against the black used for points which are within tolerance. The programme can also be arranged to stop the machine if tolerances are exceeded, thus allowing the operator to examine the area carefully. Due allowance can be made for the thickness of sheet metal between the two halves of the die. The process is completed very quickly, and loft templates are not needed, thus giving a large saving in cost and time — an important factor in each case in a market where it is often their own new model.

After proving the die sets on the presses, the finished panels can again be compared with the original tape or a derivative of it, with print-out facilities exactly as the process for checking the dies. Again by previous methods, another set of templates would be needed, and these are eliminated by the Alpha machine. Normally, at this stage, the joint faces are checked prior to welding up a prototype body. This in its complete form usually occupies the complete capacity of the machine and its system, but again scanning and contouring with the master tape is automatic for virtually the whole of the contour. Programmed comparison between a pre-prepared tape can also be undertaken for the steering, suspension and engine-mounting points together with the door, boot and bonnet apertures. However, this is often done on a semi-automatic or near-manual basis as experience of the product is often an important factor in the measurement of these points. The consumer is accustomed to seeing provision made for the adjustment of the position of doors and boot lid, but, where these advanced techniques are used, it is common to see little or no provision made for adjustment, as the dies and subsequent body parts are made to such a degree of precision that a proper fit and interchangeability are guaranteed.

Checking of Unit Parts

At a later stage in manufacture, the machine is available to check production versions of the body parts and complete bodies and to run checks periodically on the dies to ascertain the amount of wear which has taken place. A print-out is given of the comparison between the original tape and the present state of the die with a pre-determined tolerance factor to show areas which are outside the permissible amount of wear. Again, to consider the original model of the body usually made by hand, there are many small- or medium-sized parts to fit around it. These models can often be made by the Alpha machine after being presented with a tape bearing rudimentary dimensional information. The computer interpolates positions in between the points at which dimensional information has been given and mills a smooth contour in reasonably soft material for visual inspection. This technique applies to the many parts made in plastics.

Figure 6.4.
A DEA measuring machine scanning a finished car

The computer employs three memory cores, the central computer having a capacity up to 64 k bits, i.e. 64,000 individual pieces of information can be stored. Transfer of data through the system takes 1.4 μs, while a further read-only memory using magnetic cores and low-capacity scratch pad memory is provided for short-term information storage. Up to 256 peripherals can be called up. The computer is also available for certain other machines in the DEA range.

Specialized Computation

To complete the complex it is necessary to describe the functioning of another DEA machine, the Leonardo 02. The first machine was made for the Ford company of America to solve a particular problem. Its body-styling department covers a large area, and in the main the body models are of plaster or clay. The company's policy is to design a long way ahead of requirement so that transportation of a dry, brittle and heavy model to the measuring machine presented a problem. DEA solved this in the Leonardo 02 machine (perhaps indicating a future trend) by designing a machine which can be driven to the model to be scanned under its own power, then scans the model and returns to base containing the necessary dimensional information. The machine can climb gradients of 1 in 10 and can operate over a distance of a mile from its base. Severe weight restrictions were imposed, and a particular problem of floor deformation had to be overcome. The amount of floor deformation changed as the machine head moved around in scanning, and a large quantity of mercury in the base of the machine can be automatically moved around to keep the centre of gravity of the machine in a constant position despite movement of the head. The mercury is also used to provide a reference level or horizontal datum to which the computer automatically refers and compensates for floor irregularities. Figure 6.4 shows the machine scanning a finished car. The probe deserves particular mention in that it uses a force of less than 3 gf and can therefore be used on finished work or fresh clay models without risk of damage.

Figure 6.5 shows the probe in use on a car, checking small details around one of the door apertures. It can be seen that the probe can be rotated in the vertical plane to allow the job to be approached from either side, from above or below. The computer compensates for the new position taken up by the probe tip, and manual compensation is not therefore needed. As well as providing dimensional information to the computer for processing, the probe provides information for the servo-motors driving the machine during continuous scanning from its differential transducer head. Controls for swivelling the probe head are mounted on a small portable control board which the operator carries with him. Two probes have been fitted to certain Alpha machines with two rams, enabling simultaneous work to be carried out on both sides of a body model. Certain of the computer facilities have to be shared and some duplicated in this case.

Figure 6.5.
*The probe shown
checking the details
around a car door*

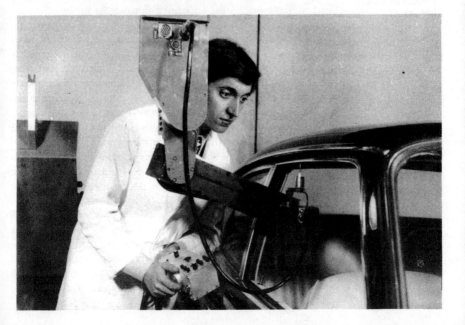

Figure 6.6.
*A television head
scanning drawings
and patterns,
mounted on the
vertical column of
a Delta 3D*

Use of Television Head

For the scanning of drawings and patterns a television head can be mounted on the vertical column of the large DEA machines; Figure 6.6 shows this for the Delta 3D machine. Telecontrol with a television screen for scanning when a tele-camera is mounted on the Alpha 3D ram is an alternative position. Figure 6.7 shows the control console for the closed-circuit television monitor which is used for operator and manual controls for the various movements of the machine. The operator is thus able to sit away from the actual work-piece or, alternatively, to work in manually inaccessible places, watching the measurements taking place and intervening manually if required.

The two large levers seen on the right and left of the console control are the servo motors driving the X- and Y-axis movements, while the lever on the right-hand side controls the Z-axis movement. The typewriter is to enable feed and speed information for the die sinker to be inserted on the tape which is being prepared while the machine is working.

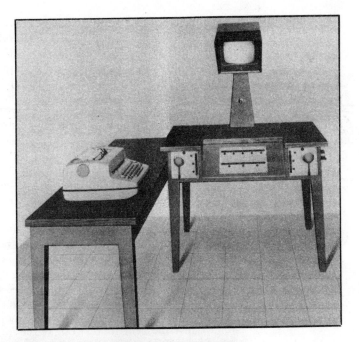

Figure 6.7.
*A control console
for a closed-circuit
television monitor;
the telecamera is
mounted on an
Alpha 3D ram*

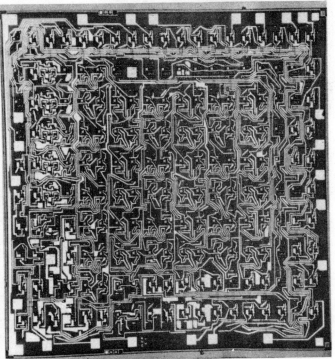

Figure 6.8.
*A photomicrograph
of a silicon 'chip'
with 187 gates*

Component Miniaturization

An example of component miniaturization is shown in Figure 6.8 which is a photo-micrograph of a single silicon 'chip' which contains 187 gates. This tiny micro-logic circuit has been developed by Ferranti Ltd, and in the form shown it is not prepared for its final use. This same basic chip has its gates connected together by an aluminium mask which is 'customized' to the individual requirements of the computation process so that the basic integrated circuit, as shown, can be adapted to a variety of needs and is known as the uncommitted logic array (ULC).

This type of circuitry has decreased the size of complicated calculators to the size of a diary, and of course, on a larger scale, machine tool and measuring machine computers now may occupy as little as one-thousandth of the volume which they would have occupied if the circuitry had employed valves.

A further, and in many cases, important result of this miniaturization has been the reduction in power consumption necessary for a given computation process. The small amount of heat dissipated by these devices ensures long life and, of course, absolute portability as dry rechargeable batteries can be used for all except the largest installations. Thus it is no wonder that the benefits that can accrue from computerized inspection are now beginning to be appreciated by progressive firms and that this feature will increase as further developments take place.

Later generations of measuring machines, made by several manufacturers, employ granite surface tables as the basis for the machines. The great stability of granite permits very high standards of accuracy to be maintained as its coefficient of expansion is low and the rust problems of iron tables do not occur. An L.K. Maxi-Check machine is shown in Figure 6.9 which embodies a black granite surface table having a guaranteed surface flatness to within 0.01 mm. The table is supported on three points which allow it to be levelled during installation irrespective of the plane of the floor.

The bridge is supported on one side by re-circulating ball bushes which run along a precision ground bar which is itself spaced from the table by adjustable V blocks. The other side of the bridge is carried on an air bearing, operating at over 40 lb/in.^{2}, the bridge being thus raised from the table by approximately 0.0025 mm. Very free movement of the bridge is thus assured, and high-quality seals are employed in the air bearings to eliminate air flow virtually and consequent local cooling of the granite (Figure 6.10).

A similar arrangement is carried within the cast-iron bridge to support the carriage and to provide Y-axis movement.

Mounted within the carriage is a rigid square-section ram

Figure 6.9.
A quality audit of Triumph TR7 body shell at British Leyland

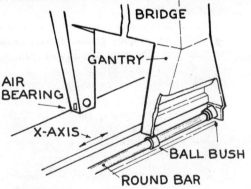

Figure 6.10.
The construction of a measuring machine to ensure freedom of movement

supported on twelve roller bearings positioned in such a way as to allow all sideways movement of the ram to be eliminated by adjustment. A smooth movement is thus also assured for the Z axis.

Each axis can be locked using vacuum clamps which incorporate fine positional adjustment from pulse motors, and in this way the ram can be 'inched' into any position over the table. The position of the ram is indicated by three digital read-out units taking their information from linear inductosyns mounted on each axis and scanned by a reader head.

Figure 6.11 shows a Maxi-Check machine equipped with a model 390 teletype, a tape reader and a Hewlett-Packard mini-computer. Instructions for automatic inspection are given to the machine by this equipment, and the various motorized movements permit them to be followed; a print-out of results can be automatically made whilst the process is continuing.

The machine can be instructed directly from a handset carried around by the operator, directly from the teletype or from a punched tape.

Tapes are made either on the teletype or in an inspection planning office in which case they may also give the operator guidance in the use of inspection fixtures and probes, the information being printed by the teletype as the tape is read. The computer takes care of all calculations which are necessary during the inspection process and also instructs axis movements and controls the probe.

Machines of this type have been made in lengths up to 30 ft long; one of these has regularly completed quality audits on sample car bodies in times around 2 hr, an operation which previously took 3½ days; this gives a convincing demonstration of the power of CNC as applied to inspection machines.

Future developments from the L.K. Tool Co. Ltd are expected to include machines constructed entirely of granite with the consequent possibility of standards room accuracies on the shop floor.

Figure 6.11.
An L.K. Maxi-Check machine equipped with a mini-computer, a tape punch and a teletype

On-machine Inspection

Frequently, in the past, this has meant in-process gauging which can really be considered an aid during manufacture, but there are many 'skin' components in the aircraft industry which justify inspection on the machine before removal. For instance, many wing surfaces are machined from thick slabs of aluminium alloy to produce the correct contour on the outer surface with a similar contour internally but with stiffening ribs machined integrally. Components such as these demand very careful setting on an inspection table or measuring machine owing to their flexibility, and as this is a very time-consuming operation a case can be made for their inspection on, or by, the manufacturing machine.

A number of objections can be raised to this method which include doubts about the fundamental validity of using the manufacturing machine to inspect its own work and the high cost of having an expensive machine stopped whilst inspection takes place.

Currently under development by the L.K. Tool Co. Ltd is a portable measuring head containing a universal probe which can be fastened to the ram of a large milling machine such as is usually used for these components and which sends a modulated radio signal back to a mini-computer and tape reader, enabling a direct comparison to be made between an inspection tape and the component. The normal machine movements are used to carry the probe around the component.

In principle it would seem to be correct to measure the 'first off' of any group of parts on a separate measuring machine as a manufacturing tape proving exercise, but subsequent specimens could be measured in the way described provided that periodic calibration checks were made on the machine. As has been mentioned elsewhere, inaccuracy information can be stored in the computer, and compensation for these inaccuracies can be made automatically during inspection. A further safeguard against manufacturing errors being duplicated during inspection could be to mount the measuring head in a different position to that taken by the cutters, thus using different parts of slides and screws during the measuring process to those used during manufacture.

7.
Automatic & Multi-dimension Inspection Machines

Introduction

In spite of the developments in gauging techniques the high-speed production of machine tools and production methods have tended to outpace methods of inspection. Thus in many cases the inspection of mass-produced components has involved a percentage check only, and inevitably some incorrect parts are thereby passed. While providing a reliable method of controlling parts during manufacture, quality control does not eliminate faulty parts from the bulk produced.

The only fully reliable method of inspection, therefore, is 100% check of all dimensions on all parts produced, and machines are available for full, semi-automatic or hand operation, permitting full inspection at low labour costs. As many as thirty dimensions can be checked simultaneously, with the machines arranged to check not only external and internal diameters but also taper, eccentricity, squareness, ovality or lobing. Counters can be provided, and on automatic machines the components can be deflected into different containers or can be graded for selective assembly.

Inspection machines can be divided into two general types: those for multi-dimension inspection of complex units such as pistons, crank-shafts and, say, connecting rods, the components being hand located for inspection; more simple units such as ball and roller bearings, piston rings, and gudgeon pins, where only a few dimensions are checked (these lend themselves to hopper feeding to the gauging points and thence are deflected to a track leading to containers). With hopper feeding one operator can attend to four machines, maintaining output up to 4,000 parts per hour per machine. With magazine feed, the output can be up to the maximum that the operator can load the components, usually about 2,500 per hour. An electronic sorting machine will check and grade simple parts at speeds of up to 10,000

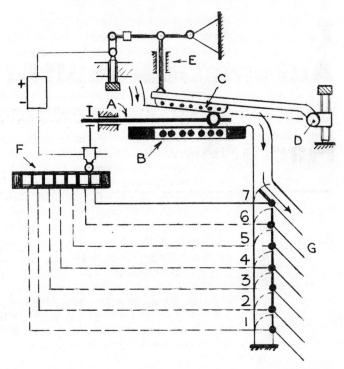

Figure 7.1.
A diagram of the mechanism for sorting and checking balls and rollers:

A *feed slide*
B *gauging table*
C *straight edge*
D *pivot*
E *electric gauge head*
F *flat commutator*
G *sorting chutes*

per hour, accuracy of inspection being in the region of 0.001 mm. In all cases the cost of inspection on these machines is very low.

Inspection and Sorting Machine

Figure 7.1 shows the mechanism employed on a machine for checking and sorting balls and rollers to an accuracy of 0.0001 mm. The components are loaded into a hopper and are transferred by a feed slide A onto a gauging table B. Above this is a straight edge C which is pivoted at D and which can be adjusted for inclination to form, in effect, a taper gap gauge. The end of the straight edge remote from the pivot makes contact with the vertical spindle of an electric gauge head E. Attached to the feed slide is a contact arm which moves over a flat commutator F, whereby electrical circuits are controlled to open and close the shutters of the sorting chutes G. The workpiece, carried forward by the feed slide, touches the straight edge at a point in the travel which depends upon the size of the piece; as soon as the straight edge is lifted slightly, the electrical contacts in the gauge head are closed, and the shutter corresponding to the associated commutator segment is allowed to open so that the work is discharged into a grading box. A system of signal lamps indicates the manner in which grading is progressing, and change-over of the machine from one size or type of workpiece can be completed in 10 min. Up to seven classifications can be obtained, and the size step between the groups may range from 0.0005 mm to 0.1 mm.

The output of the machine is adjustable from 400 to 7,000 pieces per hour, according to the size being handled.

Four-diameter Inspection Machine

Figure 7.2(a) shows a Censor eight-station machine for checking small components of the type shown in Figure 7.2(b), where four diameters of X are inspected. There are two vertical gauging heads — one at A, while that at B is supported by an anvil C. The components are loaded into the magazine D, whence they are delivered into openings in the indexing carrier E. During indexing the components are supported by the guide track F, which has openings at the gauging positions for the anvils.

The disc G is mounted at the lower end of the carrier shaft indexed by mechanism in the base to bring components to the gauging positions. Mounted on G is a pair of condensers H for each opening in the carrier E. At the end of each indexing motion, the gauging heads are connected to the electrical supply by a switch operated by the cam

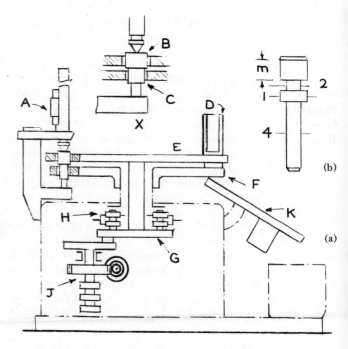

Figure 7.2.
A diagram of an eight-station machine for checking diameters:

A & B *vertical gauging heads*
C *anvil*
D *magazine*
E *indexing carrier*
F *guide track*
G *disc*
H *condensers*
(By courtesy of Censor Ltd.)

shaft J driven from the index mechanism. Simultaneously, the gauging heads are connected electrically to the condensers by sliding contacts.

When the dimension being checked by a gauging head is within tolerance, the movement of the measuring plunger causes a pivoted electrical contact to be held clear between two fixed contacts. Consequently, neither condenser of the pair associated with the particular workpiece is energized. However, when the workpiece dimension exceeds or is below tolerance, the pivoted contacts touches one or other of the fixed contacts. As a result, the corresponding condenser of the pair is charged electrically to the same potential as that of the supply.

During subsequent indexing movements, the condenser is maintained in the charged condition until it makes contact with another pair of sliding contacts at the unloading station. Current discharged by the condenser then causes one of two pivoted gates in the chute K to be set, so that when the work if unloaded from carrier E it is directed into a container. These are provided at both sides of the chute, one to receive rectifiable work and the other to reject. At the next indexing cycle, the pivoted gate in the chute is swung to its original position by the action of another switch operated by the cam shaft J. A third cam on this shaft operates a switch for controlling solenoids on the gauging heads, which are used in certain instances to assist in locating the work.

When the work dimensions are within the tolerances, so that neither condenser is energized, the gates in the discharge chute are held clear, and the component is delivered into a third centrally placed container.

Sorting and Gauging Bearing Balls

Another Censor machine (Figure 7.3) will handle these units up to 4,500 per hour; these are sorted into ten groups within diameter tolerance, the difference in group diameter being varied in steps from 0.0002 to 0.004 mm. From the hopper unit A, the balls are fed into a feed tube F and are transferred from the lower end to the gauging position, in turn, by a pusher blade attached to the slide B at the left-hand side of the measuring head C. The slides B and D are traversed horizontally by separate cams E, through connecting rods, and at the beginning of the gauging cycle they are positioned at the left-hand end of their strokes. The first ball in the tube F then falls into a slot at the right-hand end of the pusher blade to be supported by an anvil. Motion is now imparted to the slide B to the right, and, when the ball has been brought to the centre of the measuring head, the traverse speed is reduced.

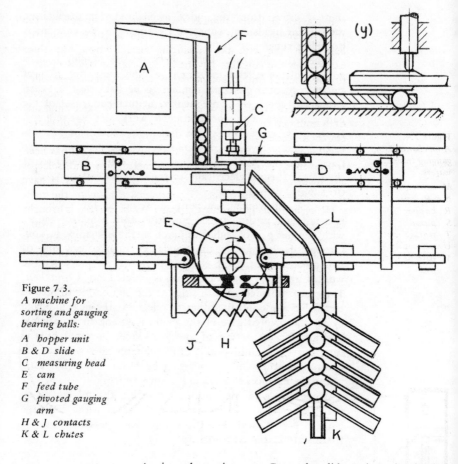

Figure 7.3.
*A machine for
sorting and gauging
bearing balls:*

A   *hopper unit*
B & D   *slide*
C   *measuring head*
E   *cam*
F   *feed tube*
G   *pivoted gauging
      arm*
H & J   *contacts*
K & L   *chutes*

A pivoted gauging arm G on the slide D is spring loaded upwards so that its left-hand end is held in contact with the plunger in the measuring head C. During the high-speed movement of the slide B, the slide D is stationary, and raised upper and lower contact surfaces at the end of the arm G remote from the pivot (Figure 7.3(y)) are positioned at the left-hand side of the measuring head. Thus the arm G is tilted so that the lower contact surface is held clear of the ball.

When the speed of slide B is reduced, that of slide D is increased to cause the gauging arm G to be swung downwards by the action of the plunger and the incline at the right-hand end of the upper contact surface. This brings the lower surface into contact with the ball, while a third cam on shaft E causes the electrical contacts H to be closed momentarily to extinguish the signal lamp. Slide D which is still traversing faster than B causes the ball to roll on the anvil by action of the lower contact surface of G. When the ball is in line with the measuring head plunger, the third cam causes the contacts J to be closed. Current is then passed to the control unit for a period of 10 ms (this is controlled by a condenser), and the signal lamp now lights up to indicate the diameter classification of the ball. At the same time one of a number of selector switches associated with discharge chutes K is energized to open a gate.

The control system is such that, after the signal has been transmitted, operation of other signal lamps and selector switches for the chutes is prevented until contacts H have been closed. Slides B and D continue to move to the right until an inclined part on the upper contact surface is brought into engagement with the plunger, so that the arm G is swung upwards to bring the lower surface clear of the ball. Slide B now increases speed to discharge the ball into chute L, before passing into the container by way of the gate.

Figure 7.4.
A mechanism for gauging transistor wafers:

M *carrier*
N & P *arms*
R *hopper*
S *pump*
T *storage unit*
U *wiper pad*
V *path of wiper pad*

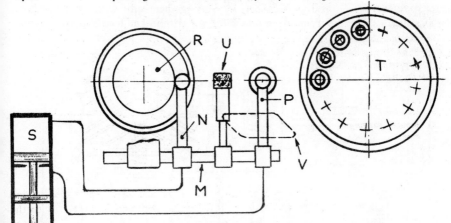

Gauging Transistor Wafers

The machine shown in Figure 7.4 will check wafers at a rate of 3,600 per hour. The cycle commences with carrier M at left-hand position with vacuum cups mounted on arms N and P and positioned above the discharge rail of the hopper R and anvil for the inductance-measuring head. Vacuum is applied by pump S so that a wafer just checked is removed from the anvil and so that a fresh wafer is picked up. As the slide is traversed to the right, the pump is reversed to deliver pressure air to the vacuum cups. In this way a fresh wafer is discharged onto the anvil, and the other wafer is passed into a compartment in the storage unit T. A wiper pad U follows the path indicated by V to remove any dust during movement of the slide.

When the vacuum cup on the arm N has been brought clear of the anvil, the slide is caused to dwell, and the plunger in the measuring head is moved down by compressed air for gauging. The plunger is then raised, and the slide is moved to the left to bring the arms N and P to the position shown. The storage unit T rotates continually by a drive from a motor clutch. When a wafer is measured, one of thirteen lamps in the storage unit is illuminated, depending on thickness. As soon as a hole in the unit T has been brought into line with the lamp, a photo-transistor head is energized, with the result that the clutch is disengaged and a brake is applied to bring the storage container for the appropriate thickness grade to rest below the vacuum cup on the arm P. At the beginning of the next measuring cycle, a cam-operated contact re-sets the grading circuit in the indicator to zero and energizes the clutch to start the drive.

Russian Inspection System

Figure 7.5 shows the mechanism of a machine for measuring bearing balls and for sorting them into seven grades in steps varying from 0.0005 to 0.001 mm. The feed hopper A termin-

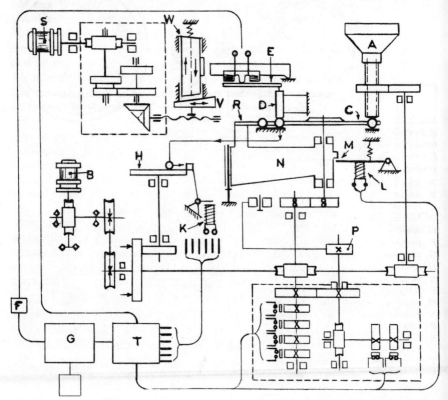

ates with a curved hole in a tube driven by motor B. As the tube rotates, balls pass down to a rotating table C, driven at 6 rev/min. A row of twenty holes near the table edge is carried past the tube outlet, and a ball drops into each hole as it passes. Each ball is then carried so that it passes between fixed lower and moving upper measuring contacts. The upper contact is carried on a floating arm D so that it is raised by a ball passing underneath, the movement being transmitted to a diamond projecting from the lever E on a pick-up head.

The lever is connected to the pivoted armature of an inductive-type measuring unit, with two coils between which the armature is pivoted on a knife edge. These coils are connected in a differential circuit which forms part of a measuring bridge with a transformer F fed from an acoustic generator in the amplifier unit G. Displacement of the diamond point causes a variation in the air gap between the armature and the coils, resulting in a change of voltage in the measuring bridge imput. This unbalance voltage is amplified and fed to a circuit containing seven relays corresponding to the classification grades.

All the relays receive the voltage, but they are also fed with support voltages of a different magnitude for each and of a reverse polarity to that of the unbalance voltage. A number of relays is thus operated, according to the signal strength, and the last to operate completes a circuit to one of seven solenoids in a centrifugal sorting arrangement in the machine base. After passing the measuring position each ball falls through a slot onto the rotating plate H of the sorting mechanism, driven by the small motor.

The ball is thrown outwards by centrifugal force so that it is retained by the rim to rotate with the plate. Surrounding the plate is a housing with seven pawls each connected to a solenoid, as at K, to move a pawl inwards above the plate, to engage a ball and to direct it down a chute into one of seven drawers.

For calibration, a switch is actuated by a timer to energize a solenoid L, whereby a locating plunger M is withdrawn from engagement with a pivoted bracket N, which carries the bearings for table C. The bracket is next moved sideways by a cam P, driven from a small gear box, so that the table is displaced sideways from the measuring position. The gear for table C is also disengaged, and its movement is stopped. At the same time, a master ball of known diameter, in a holder R attached to a bracket N, is moved into the measuring position.

The measuring bridge is now connected to a servo-motor S, through the amplifier G and a regulating unit T, and, if the

Figure 7.5. left
A Russian
inspection system
for measuring and
grading:

A *feed hopper*
B *motor*
C *rotating table*
D *floating arm*
E *lever*
F *transformer*
G *amplifier*
H *rotating plate*
K *housing*
L *solenoid*
M *locating plunger*
N *pivoted bracket*
P *cam*
R *holder*
S *servo-motor*
T *regulating unit*
V *wedge*
W *pillar*

machine setting has not changed or the measuring contacts worn, the motor remains stationary. If there is an output from the measuring bridge or wear on the contacts, the motor is driven, and movement is given by bevel gears to a screw, whereby a wedge V is displaced to give vertical adjustment of the pillar W. This supports one end of the pick-up head containing the inductive coils and the armature connected to the lever E.

The polarity of the bridge signal determines the direction of rotation of the servo-motor and pillar W. Rotation of cam P is now completed, so that the bracket N is returned, withdrawing the master ball and returning the rotary table to its original position. The complete calibration cycle occupies 1 min.

Sigma Multi-Dimension Inspection Machines

Figure 7.6.
A multi-inspection machine with turret and comparator (By courtesy of Sigma Instruments Co. Ltd.)

Figure 7.6(a) shows a mechanical system for inspection of the unit shown near the scale in Figure 7.6(b). There are ten dimensions to be checked by a turret arrangement in conjunction with a comparator head mounted on a sloping stand. The scale carries a drawing of the component, and all necessary dimensions are numbered for reference purposes, while the tolerances for the various dimensions are indicated by coloured bands. Four types of fixtures as shown in Figure 7.7 cover all requirements for external and internal diameter, external lengths and internal depths. They are adapted to inspection of a particular dimension by fitting

(a)

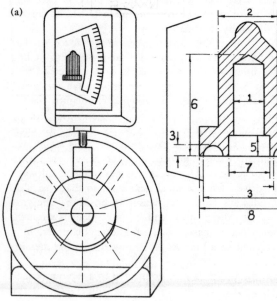

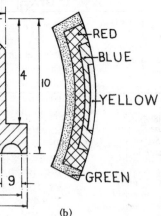

(b)

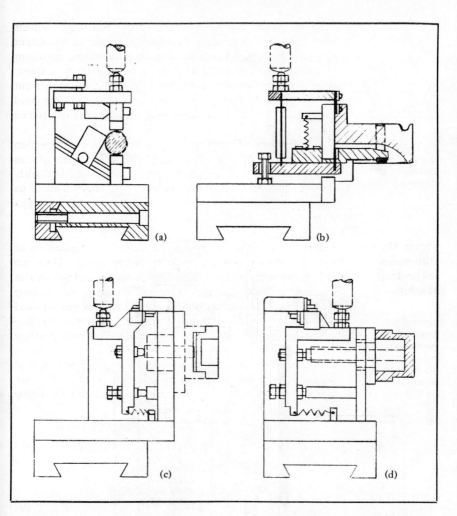

Figure 7.7.
The fixtures used on a Sigma inspection machine

small measuring anvils, and the fixtures all terminate with an adjustable setting screw which provides the contact between the inspection fixture and the spindle of the comparator head. Provision is made on the fixture for mounting a coloured strip to identify the fixture with the coloured tolerance band on the scale plate. This indicates the tolerance to which the part being measured on the fixture must be manufactured. A transparent numbered strip indicates the number of the dimension which the fixture is arranged to measure.

To set the instrument so that each dimension can be measured, a setting master is made (Figure 7.8). The dimensions are untoleranced and are target figures set half-way between each of the ten toleranced dimensions. The master

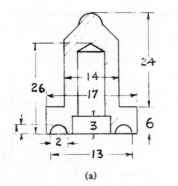
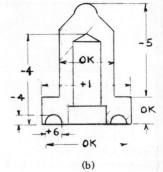

(a)   (b)

Figure 7.8.
*A setting master for
checking a ten
dimension fixture*

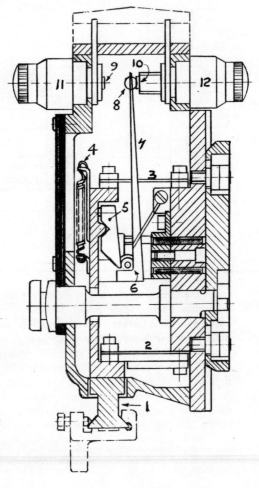

Figure 7.9.
*An electrical
signal-gauging
instrument:*

1   anvil
2 & 3   flat springs
4   coil spring
5   knife edge
6   pivoted assembly
7   arm
8   spherical contact
9 & 10   contacts
11 & 12   drums

is shown in Figure 7.8(a) and a typical result in Figure 7.8(b). The accuracy of the manufacturing equipment is checked by taking ten parts produced consecutively from the machine and by measuring them all on one dimension. The largest and smallest values are noted, and the difference in size should not exceed one-quarter of the tolerance allotted to that dimension. The turret is then rotated one station, and the next dimension is inspected, and so on. A ten-dimension fixture can be checked in less than 10 min.

Electric Signal Gauge

Another basic Sigma unit is shown in Figure 7.9. The instrument gives results to within 0.001 mm and is comprised as follows. An anvil 1 of dovetail form with contact tips is chosen to suit the work. This assembly is connected to the frame by two flat springs 2 and 3, and light contact pressure is maintained by the coil spring 4. Movement of the anvil is transmitted through the knife edge 5 to the small pivoted assembly 6 to which is fixed the arm 7. On the outer end is a spherical contact 8, the movement of which is restricted by the two contacts 9 and 10; these contacts have a screw adjustment with a precise setting by graduations on the drums 11 and 12, while the knob and locknut near the bottom provide a coarse setting.

Various types of inspection machines can be built up around this unit. One example is used to check tappets; five dimensions are sized, the indication being given by plus or minus signal lights for each dimension, while correct parts are indicated by a green light. For semi-automatic operation, the work is located on a reciprocating slide which may run continuously or operate as required from a pedal. The rate of inspection is 700 per hour.

Intricate contours such as found on turbine blades are checked on the Sigmatic electronic machine where ten dimensions on the aerofoil section of the blade require to be checked. Figure 7.10(a) shows the machine, while Figure 7.10(b) shows the dimensions checked on the blade. The conditions are different owing to the close spacing of some dimensions; therefore, between the stages of a sequence, the blade is re-set lengthways and angularly to suit the variations of twist.

For checking, a blade is mounted vertically in a fixture at the end of a vertical shaft. This shaft is moved axially to successive positions by air cylinders in two rows of five in the base of the machine. To perform the section checks, individual displacement transducers of the inductive type are associated with gauging fingers in separate groups. One group is applied horizontally over the leading and trailing edges of a blade, while another group is applied to the opposing surfaces of a blade at the mid-position across a section.

Figure 7.10(c) shows how, for successive stages of inspection, the work fixture is indexed vertically by a separate air cylinder in conjunction with stepped plates which are easily replaced. Similar arrangements provide for indexing the fixture about a vertical axis.

Control System

The cabinet shown in the illustration houses twelve electronic control units of the Sigma ten-channel type to receive electrical signals from the gauging transducers and to analyse the results. Indicating arrangements are provided which cover all the dimensions checked on a single section of blade, the section being indicated by a digital display; in operation, these dimensions are checked in turn, as shown by lamps. If a dimension is outside the pre-set tolerance, the equipment will not advance to the next check in the sequence. The indicator lamp associated with the faulty dimension will remain illuminated, and the error can then be read on an edge-type meter. A rotary switch provides for setting the number of sections at which the checks are to be made, and the unit houses the solid-state electronic logic equipment, whereby the operating sequence of the machine is controlled. Output rate for checks on ten sections is 150 per hour.

Figure 7.10.
(a) *A Sigmatic electronic machine for checking aerofoil sections:*
L.E. lateral extension
T.E. torsional extension

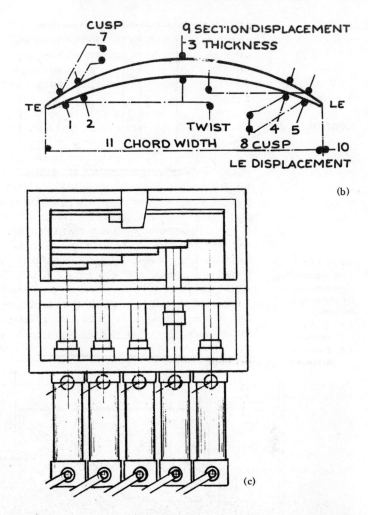

(b)

(c)

Inspection and Automatic Sorting System

Difficulties experienced with conventional electro-mechanical systems can be overcome by a combination unit employing an extremely sensitive mechanical gauge, the contacts of which are required to control only the power amplifiers (Figure 7.11). The actual electrical power handled by the contact mechanism is 0.03 W, while the power required for the whole equipment, comprising sorting, signalling and counting, is 150 W. A feature of the electronic amplifier is control of the delay or lag time of the sorting means. When the workpiece has been presented to the gauge head A, measurement takes place by displacement of the gauging spindle. The gauge contacts set the sorting gates B and deflectors D to pass the part into the category decided by the

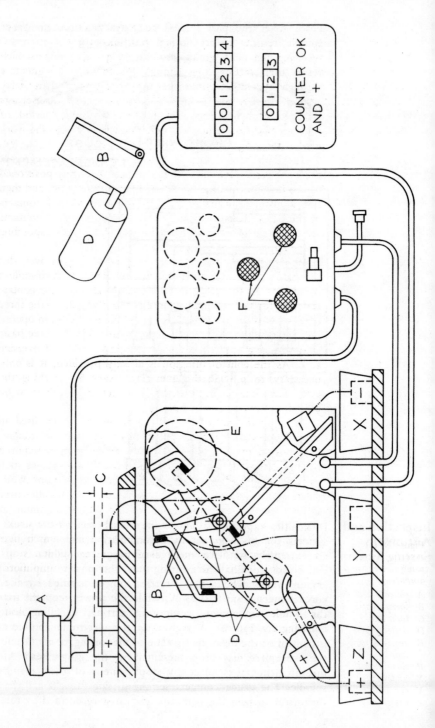

head. After the part leaves the gauge position, the gauge spindle returns to its normal position in preparation for receiving the next workpiece, but the gates or sorters must retain their setting long enough for the part to gravitate through the system into the proper category. This delay system consists of a simple and accurate condenser-charge arrangement which operates for an adjustable period of time varying from zero to a maximum of 2 s after the work has left the gauge spindle.

The limit lines C may be positioned in order to segregate the work parts three ways. The gauge head is roughly positioned to the low-limit master block on its mounting stand, and then fine adjustment is made to cause the first pair of contacts in the gauge head to close at this point. The upper contacts are then adjusted to close by using gauge blocks corresponding to the upper limit of the workpiece.

At this point the electronic gauge control takes over, this being of a plug-in type which will give upwards of ten million operations. As the relays in the control unit supply the control function for the coloured signal lights F, the next of the three basic units is the classifier. This is a sorting mechanism operating electrically to shunt the gauge parts into the three basic categories, e.g. undersize at X, acceptable at Y and oversize at Z. As the control function is already provided, it is only necessary to provide the operating means for working the gates according to the category in which the work is to be placed.

In some alternative designs solenoids have been used in conjunction with links and pivot arms to transfer the straight-line motion of the solenoid to the rotary motion required by the gate. There are, however, certain disadvantages such as rapid wear and noisy action of the link mechanism. While solenoids are capable of reliable operation, the torque curve is incorrect for this operation, for the pull is minimum at the start, increasing to maximum near the end of the stroke. This is directly opposite to the desired gate action as it gives a sluggish start with noisy wear-producing sudden stops.

For the design under consideration, then, a torque motor was developed, the characteristics of which are suited to requirements. As the rotor shaft motion is correct, the gate is fastened directly on the motor shaft, and it is possible to arrange for the maximum torque position to take place at the beginning of the stroke with minimum torque at the end. The use of Alnico rotors eliminates brushes so that the life of the sorting mechanism is practically limitless.

There is a small air-circulating unit E to direct heat away from the gauging head and to lead to stability of the equip-

Figure 7.11. left
A sorting system using an electronic amplifier:

A   *gauge head*
B   *sorting gates*
C   *limit lines*
D   *deflectors*
E   *air-circulating unit*
F   *signal lights*
X   *undersize*
Y   *acceptable*
Z   *oversize*

ment during use; covering the inner passageway with plastic pads ensures delicate handling of highly finished parts. In addition, the design of the gating system is such that it is impossible to pass work into the acceptable category without operation of the gauge head. Thus, in the event of an operator's mistake, all parts that might enter the chute without going through the gauge are rejected into the undersize compartment.

Figure 7.12 shows another application where the machine provides automatic checking of the thickness of piston rings. The apparatus is designed to be fed by the ejector of the grinder B. As the rings A enter the guide track they are fed through a cleaning bath C; then they are air dried. Next, they pass the gauging head D which measures the thickness and automatically sets the sorting gates either to accept or reject a ring. If acceptable, the gates E are positioned to drop it on a conveyor belt F which moves it to a packaging department. If rejected, it is dropped into either an undersize G or an oversize container H. Oversize rings are fed back into the machine for a second pass and are reclaimed, while undersize are either scrapped or classified for smaller-size pieces. A counter unit is shown at K.

Figure 7.12.
A machine for automatic checking of piston rings:

A   *rings*
B   *grinder*
C   *cleaning bath*
D   *gauging head*
E   *gates*
F   *conveyor belt*
G   *undersize container*
H   *oversize container*

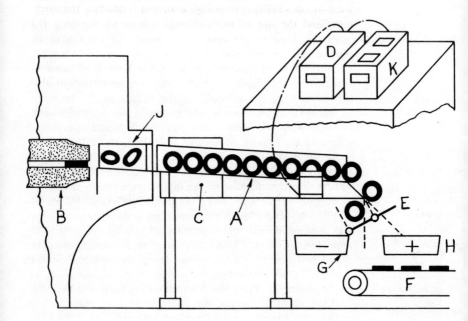

8.
In-Process Measurement & Control

Introduction

The trend in the control of workpiece accuracy is towards the use of instruments which indicate measurement during the cutting process; the instruments are built in with the machine cycle so that the operation ceases as soon as final size is reached. The earliest application of these instruments were on grinding machines where accurate sizing is more important than on most other machine tools, but recent developments have shown the value of automatic sizing on other metal-cutting machines, e.g. on lathes, where readings of size can be shown for both sliding and surfacing motions.

The earliest instruments applied to grinding machines were of the caliper type with a diamond contacting the workpiece and the size of work reduction being shown on a dial indicator. While more elaborate instruments will be described, most of the simpler instruments are of the comparator type, i.e. they are initially set from an accurate work sample and then indicate any deviation of the workpiece from this standard.

Instruments of this type must satisfy some or all of the following conditions.

(1) They must be capable of checking the work for true cylindrical shape.

(2) They should be unaffected by vibrations from the machine.

(3) They should be simple to manipulate and easy to re-set to a different diameter.

(4) The right measuring position should be automatically assumed after re-setting.

(5) There should be no damage to the work through sliding of the contact member.

(6) The instrument must be easily attached to the grinding machine.

The use of these instruments ensures that the work produced will be within the range of high-grade fits, and, if there is any deviation, it will increase as the diameter of the work increases. For this reason it is advisable to move the theoretical size of the sample for smaller diameters from the middle of the tolerance field to the minus side.

Time-sizing Devices

In most grinding operations a period of dwell is required for sparking out, and in time sizing a mechanism is fitted to control the time of dwell after the wheel has been fed against a dead stop. The time, usually between 3 and 80 s, is determined by the operator to suit the work. After a preliminary setting on a sample component the machine will grind the work to size until appreciable wheel wear has taken place. Compensation is then made, and the operation is continued. There is no gauging with the work and no contact so that splined shafts can be ground as easily as plain ones. Tolerances of from 0.007 to 0.012 mm can be maintained.

Sizing with Machine Control

The logical development was to take all responsibility for measurement from the operator and to stop the machine when the size was reached. One method is to use a photo-electric cell to interrupt a ray of light inside a casing by a moving pointer. The interruption takes place when the work size is reached. This stops the machine in-feed of the grinding wheel, while a solenoid in the control box brings the whole machine to a stand-still with the wheel head away from the workpiece.

An example of a caliper gauge is shown in Figure 8.1; this is fixed to the guard over the grinding wheel by a torsion device connecting to the caliper gauge. The accuracy of drop-arm calipers is likely to be less than that of bed-mounted gauges because they are more susceptible to vibration, sideways flexure and thermal distortion. As shown, two contact points with carbide tips locate on the work on the underside while the measuring point rests on the top position. As grinding proceeds, the downward movement of the vertical rod causes, firstly, the in-feed of the wheel head to be slowed down and finally, when the size is reached, to stop the machine.

Control by Snap Switch

Developments have included the use of ultra-high-precision snap-action switches which are operated directly as a result of size variations in a workpiece, and it is not necessary to make any provision for magnification. A switch used on sizing equipment by Vibro-Meter S.A. Fribourg, Switzerland, is shown in Figure 8.2(a), where it can be seen that it bears a general resemblance to the micro-switch used on many types of equipment. The design ensures a repeatability of the order of 0.0002 mm. The contact is carried at the end of

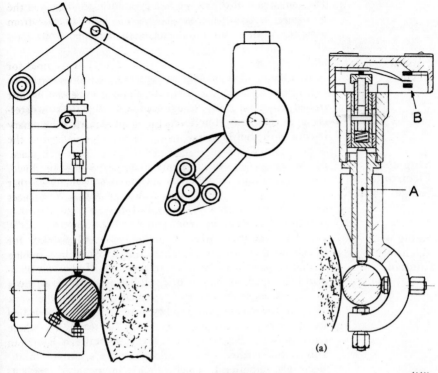

(a)

Figure 8.1.
*An in-process sizing
with caliper gauge*

Figure 8.2.
*Sizing equipment
relying on a snap-
action switch:*

*A   sensing plunger
B   switch
(By courtesy of
Vibro-meter Co.,
Switzerland)*   **(b)**

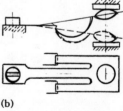

a three-leaf spring, the shape being shown in the plan view.
For attachment of the leaf spring to the body of the switch
a screw passes through a hole at the free end of the central
element. At the sides there is a shorter leaf which is permanently bowed and is hooked to an anchor point which is arranged
to project from the switch body.

In its 'idle' state the spring takes up the position shown by
the solid line. When pressure is applied to the spring by a push
rod in the direction shown by the arrow, the spring is deflected
continuously until a point is reached at which it changes its
position very rapidly to that shown by the broken line. The
time required to complete the change in position is only
1.5 ms, which is followed by a rebounding period of 3.5 ms.

The application of the switch to a cylindrical grinding is shown in Figure 8.2(a); the sensing plunger A is in direct contact with the work at the lower end, with the upper end acting directly on the sliding plunger which actuates the switch indicated at B.

By combining a number of switches, each set to work on a different work size, a complete grinding cycle can be controlled, comprising, say, coarse feed, a dwell period to correct errors in circularity, fine feed movement and dwell for sparking out to the finished size.

Stop-cote Electronic Gauge

All sizing gauges which contact the work operate under arduous conditions and should not embody delicate mechanical amplification; they should also be waterproof. A Wheatstone bridge circuit (Figure 8.3) supplied from a high-frequency generator J is made up from two fixed condensers A and B and the gauge itself, which contains two fixed plates C and D and an intermediate movable plate E. When E is exactly central, capacitances between it and each of the two outer plates are equal, so that the bridge is balanced. This is arranged to occur when the component has reached size.

Until the component reaches size, the extra amount of stock causes plate E to be nearer C, thus increasing the capacitance between the two and reducing it between E and D. This out-of-balance of the bridge creates a current in the electronic amplifier F which, after being detected, prevents a current flowing in the output stage of the amplifier. When the component has reached size, the bridge is balanced, and the current flow into the amplifier ceases. Maximum current is therefore permitted in the output stage, which is arranged to operate a relay G, causing withdrawal of the grinding wheel. Since the setting control S in one arm of the bridge has a compensator H in the opposite arm, the operation of the bridge is not affected.

Any variation in temperature, humidity, frequency and voltage of the current acts symmetrically on both sides of the bridge and consequently has no influence on the reading.

Figure 8.3.
A circuit diagram for a Stop-Cote electronic gauge:

A & B *fixed condensers*
C & D *fixed plates*
E *movable plate*
F *electronic amplifier*
G *relay*
H *compensator*

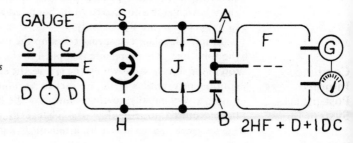

The gauge does not require any stabilization of supply voltage and allows ± 15% variation in the voltage. A bracket and two leads are required for machines with electrical control of the wheel withdrawal, while a solenoid to operate the valve control lever is necessary in the case of hydraulic control. The final setting can be done easily by means of a compensator which is not affected by backlash.

Ferranti Gauging Systems

The Pulcom systems consist of three items: a gauging unit, a drive unit and a control unit. The workpiece size is determined by the gauging unit which operates on the differential transformer system to produce an electrical signal proportional to workpiece size. Two basic types are available: contacting electronic and non-contacting air—electronic. Variants of each type cater for gauging workpieces on different types of grinding mechanism, e.g. Figure 8.4(a) shows a matching system A as the control unit of the differential type; Figure 8.4(b) indicates a transducer. The air source for internal grinding is also shown.

The gauging unit is used in conjunction with the drive unit which provides automatic advance and retraction at the appropriate stage in the machining cycle. The electrical signal from the gauging unit is amplified by the control unit which indicates component size and generates the required machine control signals. A variety of control unit types, all employing integrated-circuit electronics, provide a choice of indicator scale and output signals C.

In some machining operations it is difficult to gain access to the work to determine its size, e.g. centreless grinding where the workpiece is surrounded by the grinding wheel, regulating wheel and work rest. In such cases post-process systems can be used to measure the workpiece size immediately after discharge from the grinder and to feed back control signals to the machine's in-feed mechanisms. These signals can be used (1) for automatically adjusting for wheel wear and sources of drift as work size tends towards the tolerance limits and for stopping the machine if wheel wear is excessive and (2) for stopping the machine if a specified number of components are out of tolerance consecutively. When gauging can take place on the component, Figure 8.4(b) shows the application on a centreless grinder.

Combined In-process and Post-process System

With some grinding operations (Figure 8.4(c)) conditions may affect the in-process gauge if the component has to have ultra-high stability, e.g. for bearing manufacture. By combining both in-process and post-process gauging, the post-process gauge, which measures in relatively stable conditions, is used to modify the zero point of the in-process system so that high-stability control is achieved. Figure 8.4(d) shows how the gauging heads can be mounted in groups to check several diameters at once. The heads are of the air-electronic type and are useful for gauging crank shafts, for example, and similar work where width is very limited. Relative to two 'V' contact points an air jet is monitored by an air-electronic transducer, providing an electrical signal proportional to workpiece size.

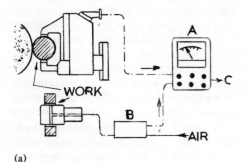

(a)

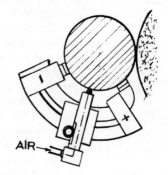

(b)

Figure 8.4.
A Pulcom differ-
ential transformer
system:

(a)

A *matching system*
B *gauging unit*
C *output signals*

(c)

A *in-process*
    control unit
B *gauging unit*
C *post-process*
    gauging unit
D *post-process*
    control unit
E *workpiece*
F *feedback signal*
G *machine control*
    panel
(By courtesy of
Ferranti Ltd.)

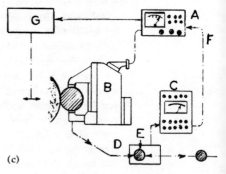

(c)

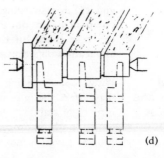

(d)

Automatic Sizing of Bores

When grinding the external surface of a shaft there is a line contact between the wheel and the outside diameter of the shaft, and ample space is available for mounting and operation of the sizing instrument. This feature does not apply to internal grinding, since in this case there is arc contact between the wheel and bore of the work, while the presence of the wheel within the bore makes it often impossible to use gauging fingers in the bore. The difficulty is greatest when the diameter of the bore is small. Also, wheels of comparatively small diameter must be used on internal grinding machines with the result that these wear more rapidly than the large wheels used for external grinding.

Electronic devices are available, but a mechanical arrangement (Figure 8.5) allows the work to be ground almost to size and the wheel to be dressed before the finishing grinding operation. The diminishing size of the wheel caused by wear and the truing operation is automatically compensated for, irrespective of the amount removed from the wheel by truing.

The mechanism consists of a bearing A fixed to the cross slide which carries the wheel head. A rod B passes through the bearing and is prevented by friction from moving axially unless deliberately displaced; it will therefore move with the wheel slide in a direction at right angles to the axis of the wheel spindle. Rod B is connected to the pointer C of a dial indicator, the pointer registering zero when the work is ground to finished size.

As each component is mounted in the machine, the wheel slide must be withdrawn a sufficient distance to permit the wheel to clear the unground bore. Rod B moves with the slide, and indicator C rotates in an anti-clockwise direction. Grinding is then begun again, and, if the grinding wheel did not wear, the finished size would be indicated by the pointer reaching zero. However, the wheel does wear and therefore, as soon as the zero position is reached, it is withdrawn and trued by diamond D, which is brought into position hydraulically

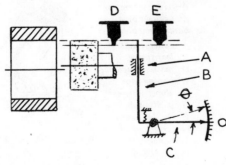

Figure 8.5.
*A mechanical
system for sizing
bores on internal
grinder:*

A *bearing*
B *rod*
C *dial indicator
  pointer*
D *diamond*
E *stop*

As the wheel is being trued, the end of rod B makes contact with stop E and is pushed back through a distance equal to that by which the diamond projects beyond the finished surface of the work. The indicator finger is thus moved in an anti-clockwise direction through the angle θ. Grinding is then continued until the indicator again reaches the zero

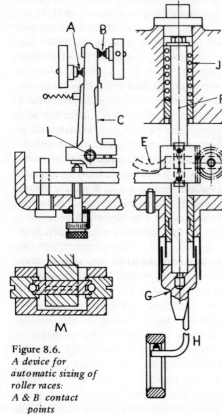

Figure 8.6.
*A device for
automatic sizing of
roller races:*

A & B contact
      points
C lever
E latch
F spindle
G sleeve
H diamond bar
J spring
L centre

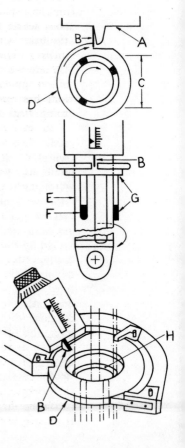

Figure 8.7.
*A sizing system for
ring gauges on a
honing machine:*

A control unit
B leaf spring
D master gauge ring
E honing tool
F abrasive stones
G plastic tubs
H carbide insert
*(By courtesy of
Micromatic Hone
Corp., U.S.A.)*

position; the wear of the wheel during the light finishing cut is not sufficient to affect work accuracy. The wheel truing stroke and return table movement to the loading position on completion of the grinding operation are both controlled by electric contacts operated by the sizing device.

Van Norman Cinophon Electronic Gauge

The device shown in Figure 8.6 is designed primarily for the automatic sizing of roller races using a diamond contact on the bore and registering the amount of material being ground out in comparison with a given size ring. When the work has been ground sufficiently to allow the previously set electric contact points in the gauge to close, these actuate relays in the circuit of the machine, thus stopping the automatic feed to the wheel and gauging the work bore to the set relation to the size ring. The mechanism comprises the spindle F and sleeve G which carries the diamond bar H. The diamond is kept in contact with the work by spring J, and, as the spindle moves upwards, when grinding takes place the point of screw contacts a plug on the base of the lever, causing it to pivot about centre L so that contact at the top of the lever makes or breaks with fixed contact points mounted in Bakelite blocks. An enlarged view is shown in Figure 8.6(M).

There are two sets of contact points A and B which have entirely different functions. The first set operates before the work has reached final size and regulates the rate of feed during final grinding. This set of points opens, breaking a circuit to the Cinephon electronic unit. The second set of points B, which close, are the actual sizing points, and these close another circuit. The moving halves of the two sets of points are carried on a lever C, actuated by the diamond arm which moves over between the fixed points carried on insulated blocks mounted in the gauge head. During the early part of grinding, lever C is at rest on the left-hand fixed point A.

The diamond arm is not directly connected to the lever but operates it only when a ball point on the arm reaches a plug on the lever. While the lever is resting on the left-hand fixed point (that is before the arm reaches it and starts to move it), the machine feeds in at a controllable fast rate, roughing the bore. As soon as the diamond arm moves the lever, the first set of contact points opens, breaking the left-hand or normally closed contacts. Through the Cinephon the rate of feed is then transferred to a second control which is slower than the previous rate, thus giving a finishing out to the workpiece.

During the time this finishing out is in operation, the lever is moving over between the left- and right-hand points, and, when it reaches the second point and closes the circuit, the in-feed is stopped altogether. This also starts the vacuum tube time relay which allows the grinding out to run out and finally, by the action of the time relay, to withdraw the wheel from the bore. A latch is provided to allow the diamond to be moved in or out of the ring; this is shown at E.

Automatic Sizing on Honing Operations

Automatic gauge ring sizing can be either electronic or pneumatic in operation. Figure 8.7 shows the first type in which E is the honing tool and D the master gauge ring. Plastic tabs G at each end of the tool wear down with the abrasive so that the outside diameter of the tabs is always exactly equal to the diameter of the bore being honed. The gauge ring is positioned directly above the work, so that the upper tabs enter the ring at the top of each stroke. The abrasive stones F do not touch the surface of the sizing ring. The inner diameter of the gauge ring equals the low limit of the desired bore size. When the diameters of the bore and the tabs equal minimum bore size, the friction between the upper tabs and gauge ring becomes great enough to cause the ring to turn slightly with the tool. As the gauge ring turns, a notch on the periphery acts on the leaf spring B, the movement causing electronic control unit A to be grounded and short circuited, thus preventing further stone expansion and ending the cycle.

Pneumatic Theory

As air travels through a pipe, a certain amount of energy is represented by its mass and velocity. If the flow is stopped suddenly, the air in the closed end of the circuit will be compressed to greater than line pressure, and a pulse travelling at high speed will be deflected towards the opposite end of the pipe. This phenomenon causes air hammering when a pipe is suddenly closed. The pulse will travel back through the pipe even if the downstream end is re-opened. Such a pulse can be used to operate a sensitive pressure switch, the

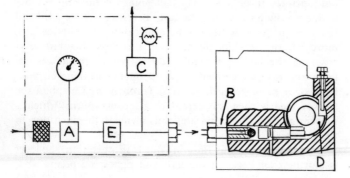

Figure 8.8.
The pneumatic theory of air sizing of gauge ring:

A   *regulator*
B   *ball-seat assembly*
C   *pressure switch*
D   *gauge ring*
E   *adjustable restriction*

speed of operation not being noticeably affected by the length of the pipe.

Figure 8.8 shows a gauge-ring-sizing design to detect the rapid movement of the gauging ring. Air flows through a regulator A at a pressure ranging from 1 to 1.4 kg/cm^2 and thence through an adjustable restriction E, a flexible line leading to the sizing bracket. From there it goes past a ball-seat assembly B and around the ball between a plunger and the bracket wall to the atmosphere by way of a release hole. Clearance between the plunger and wall is such that a pressure of 0.6 kg/cm^2 is maintained behind the plunger. A sensitive pressure switch C, connected to the feed-stop control of the machine, is set to function at a pressure of 0.7 kg/cm^2.

When the tabs on the ends of the abrasive sticks contact the gauge ring D, it is caused to rotate with the tool, to contact the valve plunger and to move it against the air pressure, without altering the flow rate, until the ball is pressed against the seat. The impulse formed at the instant of closure is reflected back through the air hose to the pressure switch, causing it to operate. Differential action of the switch provides the time necessary to actuate a relay to interrupt the feed-out movement of the abrasive sticks. Although the ball may be on its seat for as short a time as 0.001 s, the switch may remain closed for as long as 0.1 s. The purpose of the gauge ring stop is to limit the float of the ring. The run-out time is determined by an electric timer.

Auto-sizing on Surface Grinders

Systems which measure the positioning of the grinding head with respect to the work table are of little use since they do not take into account the wear of the grinding wheel. Also, systems used to monitor the face of the grinding wheel continually, using jets of air, work under difficult conditions in attempting to compensate for wheel wear. Similarly, attempts

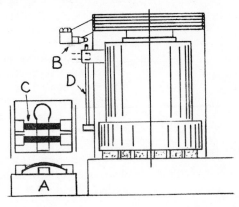

Figure 8.9.
*An automatic
sizing system for
surface grinding:*
A *insulated
    clamping block*
B *adjustable limit
    switch*
C *copper strips*
D *rod*
*(By courtesy of
Snow and Co. Ltd.,
Sheffield)*

have been made to measure the actual face of the work being ground, but, owing to the intermittent cutting action typical of reciprocating table grinding machines, the gauging head is subject to wear and vulnerable to damage. In these systems the machine frame is used as part of the gauging system. Thus, small changes in temperature cause relatively large changes in size, and corrections must be made over the working day.

A system (Figure 8.9) has been developed by Snow and Co. Ltd., Sheffield, to measure the actual distance between the grinding wheel and the surface of the moving table. A small strip of printed-circuit board (plastic material with a thin coat of copper foil bonded to one face) is held in an insulated clamping block A, so that the copper foil completes the electrical circuit to apply incremental feed to the grinding wheel. The clamping block is adjusted in height so that when the grinding wheel cuts through the copper foil, the feed is interrupted. It is simple to arrange two of these strips C with one of them 0.120 mm higher than the other. Then the machine can be arranged to grind roughly with coarse feed and to change automatically to a fine finishing feed.

At the end of the grinding cycle it is necessary for the head to be raised to clear the next unground workpiece. This is achieved automatically by a rod D sliding within a friction grip against a fixed stop until the workpiece is ground to size. The grinding head and rod then retract until an adjustable limit switch B is operated. The amount of retraction can be set for variable stock removal and takes account of all abrasive losses. By integrating the sizing head and wheel compensating arrangements with the standard controls on a surface grinding machine, the operator need only press the start button and the following sequence will take place.

(1) The machine and components in position with copper strips in the gauging head commences reciprocation with coolant supply switched on.

(2) At each table reversal a coarse automatic down-feed occurs until the first strip has been cut by the grinding wheel.

(3) A fine automatic down-feed of the wheel head occurs at each table reversal until a second strip has been cut, and under the control of an electrical timer the table continues to reciprocate for a pre-determined spark-out period.

(4) The table runs to the loading position, the coolant is switched off and the wheel head retracts ready for the next cycle.

(5) The work table is unloaded and re-loaded, the copper strips are replaced and the machine is ready for the next cycle.

9.
Digital Read-out Systems (Machine Tools)

Introduction

It is not surprising that up to recent times measuring during machining was largely restricted to grinding machines, for grinding is usually the main precision operation on shafts and spindles, and gauging equipment is not difficult to install on the machine. By comparison, sizing gauges on lathes have to contend with heavy-stock removal and thus require a greater range, while the surface texture is generally rough. Moreover, measurement in two directions is required; diameter control (surfacing) and length control (sliding); this latter is often of considerable length. The importance of some means of accurate measurement on lathes can be seen from the fact that in an aero-engine factory 40% of the machine tools were lathes, but that mistakes in turning accounted for 60% of the parts scrapped.

Ferranti Acculin Equipment

The Ferranti Acculin equipment comprises an electronic counter and two transducers, one measuring work diameter and the other length. The Ferranti moiré-fringe-measuring system is used, and Figure 9.1 shows the equipment 2, 3, 4 arranged for surfacing. The transducer is actuated by the push rod 6; the counter 10 incorporates a twin-track scale grating, a dual reading and an amplifier to provide alternative outputs for metric or inch readings.

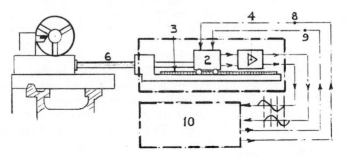

Figure 9.1.
The Ferranti Acculin measuring system:
2 reading head
3 scale grating
4 amplifier mounted on spar
6 push rod
8 & 9 lamp supply
10 counter

Figure 9.2.

*Lathe saddle move-
ment measuring by
rotary transducer:*

1 *wheel*
2 *reading head*
3 *twin-track radial
   grating*
4 *amplifier*
7 & 8 *lamp supply*
9 *counter*

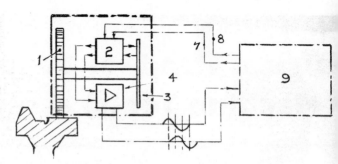

Saddle movement along the bed is measured by a friction-driven rotary transducer. The linear movement is converted to rotary via a wheel rolling along a slideway. One revolution tracks out at 254 mm. Figure 9.2 shows the wheel 1 coupled to an angular measuring system which provides alternative outputs (metric or inch). The equipment is easily mounted on any new or existing lathe and is completely protected against the ingress of cuttings or coolant.

**Counter
Cabinets**

Figure 9.3 shows the Acculin L measuring system; the diameter display sizes are on the top line and the length display below. Also on the top line are the thumb switches used to set datum on the diameter-measuring axis. Between these switches and the figures is a push button with a dual function. It causes the number set on the switches to be read into the counter memories, and it re-sets the check circuits on the diameter axis. Directly underneath is the counter re-set button for the length axis. There is also a direction sign indicator, and two fault finders are provided for each axis, one indicates a transducer fault and the other a counter cabinet fault.

The introduction into a workshop of these measuring systems can lead to great increase in productivity, especially if the read-out is coupled to pre-set interchangeable tooling. In this case the operator must be supplied with a set of numbered tools and a pre-setting tool fixture. This enables tools to be set to a standard relationship with respect to length (X axis), to diameter (Y axis) and to height (Z axis) on the fixture. Thus, when the tools in their holders are transferred to the lathe, the datum relationship is maintained. Before direct read-out of the diameter being cut can be obtained, the diameter display must be set so that, when the tool tip coincides with the centre of the lathe spindle, the display reads all-zeros. This operation is called 'setting datum', and

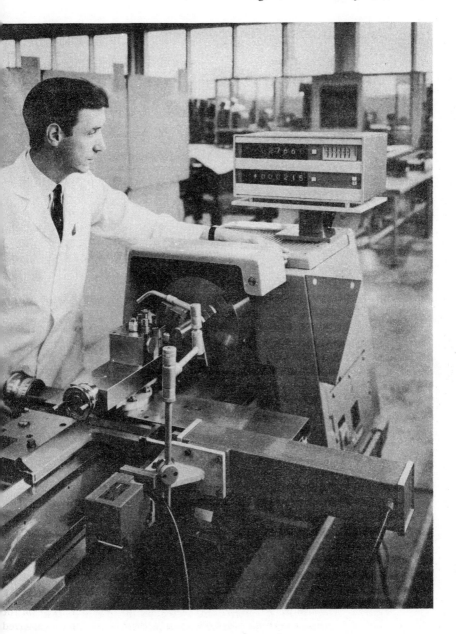

Figure 9.3.
The counter cabinet mounted on a lathe

it only needs to be carried out once a day or before measuring a component. The simplest way to do this is to take a suitable cut, to measure the diameter with a micrometer and then to pre-set this number into the diameter display.

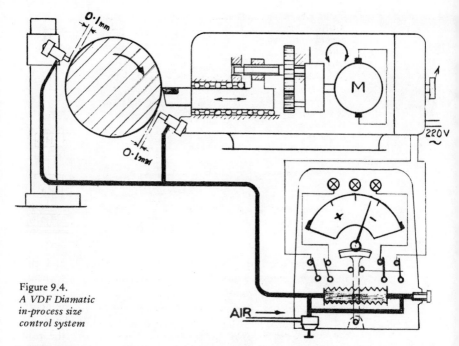

Figure 9.4.
*A VDF Diamatic
in-process size
control system*

**VDF Diamatic
Size Control
System**

This German development is shown in Figure 9.4 and incorporates a tool holder with a motor-driven fine adjustment, controlled by a two-point measuring system to maintain automatically a pre-set turning diameter on the workpiece. Adjustment of the tool position can be made in increments as small as 0.001 mm; these are shown on a combined dial indicator and control unit. The tool holder slide is traversed by a d.c. motor through a gear train, screw and nut, and speed can be varied in steps by a potentiometer to suit cutting conditions.

The control unit operates in conjunction with two air gauging heads with carbide jets which are arranged for micrometer adjustment and which do not contact the workpiece. One gauging head is mounted on the fixed portion of the tool holder below the cutting tool. The second head is supported on a column at the rear of the machine and can be adjusted as required in all planes. Both jets are of the high pressure type and are connected to the dial indicator and control unit which provides a magnification of 5,000 ×; any change in the air gap is shown on the dial. Command signals to advance, withdraw or stop the tool holder slide are transmitted by the control unit. Since high-pressure air is used, the presence of coolant does not effect the measurement, and pieces of swarf are blown clear of the jets.

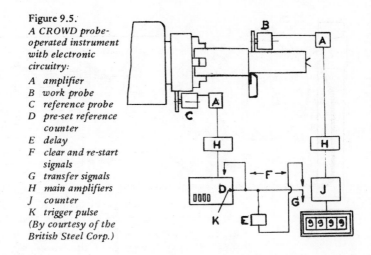

Figure 9.5.
A CROWD probe-operated instrument with electronic circuitry:

A *amplifier*
B *work probe*
C *reference probe*
D *pre-set reference counter*
E *delay*
F *clear and re-start signals*
G *transfer signals*
H *main amplifiers*
J *counter*
K *trigger pulse*
(By courtesy of the British Steel Corp.)

In operation, a workpiece is rough machined, and the diameter is measured. With feeler gauges the jets are set using the micrometer adjustments. The cutting tool is brought lightly into contact with the surface of the work, the depth of cut for finish turning is set (less than 1 mm) and the jets are advanced by the same amount. Immediately the tool starts to cut and the jets become operative, the pointer of the dial indicator is set to zero, and the electrical system of the control is switched on. Compensation can be made for errors of size resulting from tool wear.

Probe-operated Instrument

A development of the Research Department of the British Steel Corporation for measurement during machining of large components is shown in Figure 9.5. Known as CROWD (continuous read-out of work dimension), the equipment comprises two measuring probes and a cabinet housing the electronic circuitry with a digital display. The unit is spring loaded so that it may be positioned to pre-load a feeler wheel against a rotating surface to provide frictional drive for the generating mechanism. This is in the form of one fixed radial grating and the other attached to the wheel, whereby the passage of light from a gallium arsenide source is controlled. The head incorporates a small amplifier A for signals passed to the circuits in the display unit. The work probe B contacts the work, while C operates on a known diameter, e.g. that of the chuck.

The system is based on a modified ratiometer; signals from the work are being counted, and the total is expressed as a proportion of the number of signals emanating from the reference probe in the same period. Thus the system is that of

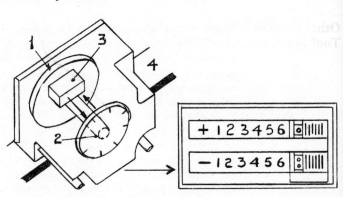

Figure 9.6.
A diagram showing twin-track grating and reading head:

1 *friction-driven wheel*
2 *twin-track grating*
3 *reading head*
4 *track*

a comparator. The sequence starts with the generation of a pulse from the pre-set reference counter D when the established total has been reached. This trigger pulse K causes the total count obtained in respect of the work probe portion of the system to be extracted, converted and used to up-date the display. Simultaneously, the pre-set counter is cleared and re-started. There is a short delay before the work probe counter is also cleared and re-started, but there is no loss of signal. These clear and re-start signals are indicated at F, the transfer signals at G with the delay at E. The main amplifiers are at H, the counter at J leading to the display.

Two good features are the ease of calibration and the avoidance of accurately matched wheels on the probes. A third feature is that the display indicated diameters are based on proportionate comparisons with the reference diameter and not on absolute measurement; the readings can be in any units metric or inch.

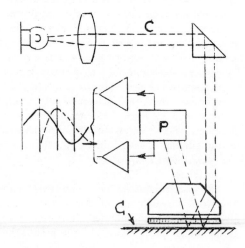

Figure 9.7.
A Ferranti Autograt system using optical and electronic units:

C *collimated light*
P *photo-cells*
G *grating*

Other Machine Tool Systems

Digital measuring systems can be applied to milling, boring and drilling machines and to a range of other machine tools. In contrast, however, with grinding machines and lathes where work size is the consideration, in general the measurement requirements on other machines are restricted to the length of slide movements or to tool depths. This is still a valuable feature, giving accurate location of tools or work, reducing the time of measurement and in many cases eliminating the use of jigs, fixtures and marking out. Transducers can measure tool or saddle movement directly without intermediate drives from lead screws or rack and pinion; the length of travel measured is limited only by the counter capacity, and for large machines travel up to 15 m can be available.

Amongst three Ferranti systems, that shown in Figure 9.6 bridges the gap between time-wasting conventional gauging techniques and high-cost NC. A typical installation comprises two friction-driven transducers and a two-axis display unit showing six digits, plus and minus, and reading to 0.01 mm. The operating principle comprises a friction-driven wheel 1 connected to a moiré fringe twin-track measuring system to generate signals proportional to tool-carriage displacement. These signals are applied to the display unit by cable.

Autograt

Figure 9.7 is an advanced design employing the latest optical and electrical units to ensure high accuracy and speed in machining operations. Directly coupled linear transducers transmit the machine movements to a six-digit display on each axis, the plus or minus signs indicating the position of the count from zero. The transducer is a self-contained unit with a rigid spar, a twin grating and a reading head, the grating G being clamped to the spar. The reading head is driven inside the spar by a 'link' which couples the machine's movements to the transducer. Relative displacement between the reading head and spar-grating assembly is sensed by a vernier fringe optical system. Either the spar or reading head can be static to suit the machine under control.

Self-contained transducers mounted on the machine convert linear motion to electrical signals which are processed and counted to provide the machine operator with a digital read-out of the exact distance travelled by the tool-slide from a selected datum.

The relative movement of the reading head with respect to the spar-grating assembly generates vernier fringes which modulate the collimated light C reflected from the spar-grating and falling onto the photo-cells P to produce a sine-cosine signal proportional to the input displacement. Move-

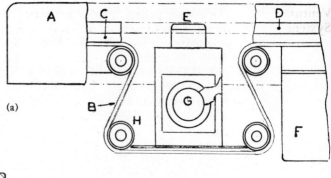

(a)

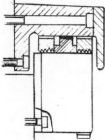

(b)

(c)

Figure 9.8.
*A seven-digit
Ferranti system for
large machines:*
(a)
A   *rigid spar*
B   *nitrile rubber
strip*
C   *aluminium
extrusion*
D   *spar grating*
E   *index grating*
F   *amplifier*
G   *lamp*
H   *rollers*
(c)
A   *photo-cells*
B   *principal focus
of lens*
C   *filament*
D   *lens*
E   *spar grating*
F   *index grating*

ments up to 3,175 mm can be measured to a resolution of 0.005 mm.

Another Ferranti system shown in Figure 9.8 (a), (b), (c) is best utilized on large-capacity machines which require good repeatability over long linear travels. Seven digits are available if required on each axis. The transducer (Figure 9.8(a)) comprises a rigid spar A with a front cover, a length of steel reflecting tape and a reading head, the tape being clamped along the length of the spar. On each axis of measurement, the reading head and the gratings are protected from the ingress of dirt and coolant by an effective sealing system, shown in Figure 9.8 (b). There is a nitrile rubber strip B which runs along the length of the spar and is fastened at the ends. As the reading head travels, the strip is pulled out of the aluminium extrusion C and is passed around the reading head and back into an extrusion at the rear.

The spar and index gratings are shown at D and E with the amplifier at F, having the cable outlet at the bottom or side if preferred. The optical block carries the lamp G. Transducers mounted on the machine convert linear motion to electrical signals which are processed and counted as described before. The basic reflecting optics (moiré fringe) are shown in Figure 9.8(c).

Venture System

The Venture System, Smith's Industries Ltd., is an incremental digital read-out system of linear measurements with noise immunity and reliable operation in harsh environments. As shown in Figure 9.9 it is made up of a signal unit or encoder and a digital display. The encoder is mounted on a fixed member of the machine whose motion causes a steel tape from C to extend and retract from the encoder; in turn, an optical-type transducer converts the tape motion into data that are translated into digital information for display. The display indicates the direction and distance to within ± 0.01 mm in any 500 mm distance, and up to 9 m can be measured.

The Smith 300 system is absolute and does not depend on counting techniques and, because of the discrete memory, the position is never lost. At the heart of the system is an absolute encoder which utilizes photo-electric sensing for the ultimate in long life. There are no contacts to wear or oxidize and no mechanical elements to wear. Long-life lamps are so derated as to extend lamp life greatly. The maximum speed of travel is 12.5 m/min, a rack providing the measuring element. A twin-axis shared display is well suited for the initial piece set-up with subsequent positioning related to the other axis; alternatively a machine with an auxiliary axis, where two given axes would not be used at the same time, is sometimes employed.

Figure 9.9.
*A Venture
incremental digital
read-out system:*

A *spring unit*
B *gear train and
   encoder*
C *steel tape*
D *measuring wheel*
*(By courtesy of
Smith's Industries
Ltd.)*

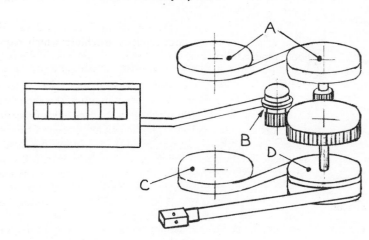

10.
Solid–State Equipment for Inspection & Measurement

Introduction

In Chapter 4 reference was made to the use of solid-state equipment on inspection machines, but the importance of this type of equipment for control and automated drive systems warrants further comment before the description of solid-state control as applied to the Landis-Lund grinding machines.

The equipment is now largely replacing other types of control and power functions and is applied to digital computers, programmers and data loggers, also to a range of high-quality and flexible analogue feedback control systems. In d.c. power systems solid-state equipment has replaced the mercury-arc rectifier and the motor-generator set. The advantages to industry include the provision of power conversion equipment of improved efficiency and versatility. In addition, advanced control systems have become practicable which have made it possible (through programmed and computer logic) to replace some of the human factor in manufacturing process control.

One of the factors responsible for this revolution is the development of the required components, namely reliable solid-state items which have important advantages as regards operation, maintenance or efficiency, compared with previous equipment. The use of solid-state equipment started with the introduction of such low-power items as diodes and transistors for digital and analogue control systems. High-power items such as power thyristors were developed on the basis of the initial work and are now being used for a variety of applications involving power requirements up to very high values.

Another factor is the extensive and increasing demand from industry for advanced automation systems as solid-state equipment is necessary to afford the performance characteristics which users demand. The three major types of equipment are (1) digital logic systems, (2) analogue feedback (linear) systems generally used to control specific functions such as speed, position, voltage or current and (3) d.c. power sources to convert a.c. power to variable-voltage d.c. for power process drive motors.

Digital Logic Systems

The development of reliable solid-state components enabled industry to take digital logic from the laboratory and to apply it for service in manufacturing operations. Relays, magnetic logic and electronic tubes have been used for sequencing, programming and computing functions, but they are not suitable for complex systems because they are short-lived, slow and expensive or because they result in high power dissipation. Solid-state digital logic components are however, so much better in all these respects that they make practicable such large and complex systems as data loggers, on-line programmers, process control computers, gauge controls, position controls and NC equipment for machine tools.

Data loggers provide the basic information from which the rules for further automation can be derived. When information is collected for constructing a mathematical model of a process, all the associated factors must be investigated, and data loggers can take information from sensing devices and can report it in a co-ordinated manner. Digital programmers are used for applications wherein production equipment can be caused to perform a series of pre-determined operations. The digital computer, referred to later, is an even more effective means of applying engineering intelligence to manufacturing operations that were previously controlled empirically. The reason is the ability of a computer to perform complex calculations quickly and to apply adaptive control and memory functions to improve operating factors.

Analogue Feedback Control Systems

Most of these systems are of the analogue type because there are inherent relationships between the elements of machines and process equipment parts which are analogue in character. Digital feedback systems have certain advantages, particularly in precision, in applications such as positional regulators, but dependence on the state of feedback information, digital-to-analogue or analogue-to-digital conversion, may be necessary at some point in the control loop. For instance, a digital-positioning system may well have to provide a d.c. output voltage at a relatively high current to drive a d.c. feed motor used for the positioning.

In regard to d.c. power sources, from 1955, germanium diodes and then later silicon diodes have been used to form d.c. power supplies with a low source impedence. Stabilization for these power supplies have been effected by power transistors and Zener control circuits for low powers, whilst high-current power supplies have utilized thyristors (silicon-controlled rectifiers). This present-day circuitry replaces the once familiar Ward-Leonard set and the thermionic rectifier, giving simpler control circuits, much less maintenance and greater efficiency and precision in regulation.

Thyristors are commonly used for field excitation supplies where a constant output is required (in terms of either voltage from a generator or speed from a motor), irrespective of variations in supply voltage or imposed load. The fast response time of these solid-state systems has led to their quick acceptance. The armature supply for the motor can be controlled in the same way.

High-power d.c. motor drives are now extending to the limits of capability of the machine tools themselves and the basic specification of such a drive (Flli Morando, Turin, Italy) follows. A d.c. main spindle motor of up to 120 hp is fitted to boring and turning mills made by this company, and wide speed variations with constant power is provided by thyristor-transistor control. The main motor has a frame size larger than normal to enable full power to be given at low speeds, and a separate a.c. ventilation motor is provided for cooling. A tachometric generator is fitted to the main motor to provide feedback information, and this in turn is processed by integrated-circuit operational amplifiers in the control loop.

The speed range is from 0 to 1,800 rev/min with a speed regulation of ± 0.5% for load variation between 5% of full load. The acceleration is timed to take between 3 and 30 s according to a pre-set control. From these developments it can be seen that the relatively low powers required for inspection and measuring machines and for control engineering can be obtained easily with even higher precision.

Digital Systems

A digital computer is made up of a number of blocks with a large number of interconnecting paths to perform various arithmetic, geometric and algebraic functions. The main signal paths are selected according to the programme or routine given to the computer and vary with the operation to be performed. Each block or group in the computer consists of a number of minor circuits which are known as elements. These generally consist of transistors or diodes which either are conducting fully or are switched off. These logic elements may have one or more imputs and one or more outputs depending on their particular function. They either accept or reject pulse information, and, when there is more than one output, the element will route the information pulse to the correct one further in the line. These outputs are often given only when a complex set of input conditions have been fulfilled.

There are four basic circuit elements which are the OR, AND, NOR and NAND (NOT AND) gates. Present-day logic design can incorporate all of these elements together with flip-flops, dividing counters, shift registers and display tube drivers. A large number of more specialized elements are also available.

Where the basic gates are concerned, a set of systematic relationships is used in the coupling of the various gates to perform specific operations, and those relationships are derived from mathematical formulae according to two-state Boolean algebra. The information is known in this state as 'binary coded' information because of the two states that the elements can be in, either conducting or non-conducting, and which are represented by either 1 or 0 respectively.

The operating principle of a NOR gate for instance is that it has no output (0) when it receives an input (1) but provides an output when it has no output. The NAND gate, on the other hand, gives no output (0) when all its gates have imputs (1), but, if any input is switched off elsewhere in the circuitry, it then gives an output.

Figure 10.1
The basic circuit elements for digital control:

X & Y switches
Z lamp

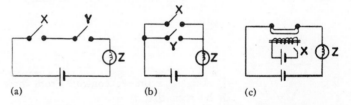

(a) (b) (c)

The AND function is an operation which may be represented by a number of electrical switches in series, as shown in Figure 10.1(a), from which it may be seen that the lamp Z will light up only if both switches X and Y are closed. This result may be written as Z = X AND Y. OR functions, on the other hand, may be illustrated by the analogous circuit in Figure 10.1(b), showing the electrical switches in parallel. It can be seen that the lamp Z is illuminated if either X or Y is closed, i.e. Z = X OR Y. Figure 10.1(c) illustrates the principle of the NOT function, showing that the lamp Z remains on as long as the relay is not energized, but, as soon as the switch X is closed, which represents a signal to the input of the gate, the lamp is extinguished. From these three basic logic functions it is possible to build up other more complex functions, such as a binary adder circuit.

Pneumatic Logic

The development of small control units to perform logic switching functions has led to their application to machine

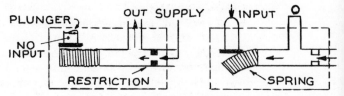

Figure 10.2
*A pneumatic micro-
switch operation*

tools and sizing devices, the subject being known as fluidics.
One system, made by Techne (Cambridge) Ltd, is based on the
use of air pressure capable of directly working robust diaphragm control valves, thus eliminating the need for pneumatic amplifiers.

A pneumatic micro-switch (Figure 10.2) shows the principle
of the unit. Air under pressure is applied, via a restriction, to
the inside of a tightly wound steel spring. When the spring is
straight, there is no loss of air, and a high output pressure
is maintained, but an output pressure of 1 kg/cm^2 will fall
to less than 0.07 kg/cm^2 for a spring movement of 0.5 cm.
The deformation of the spring is produced by the expansion of
the Neoprene rubber diaphragm which is caused to move
by an input signal of 0.28 kg/cm^2. There are two input and
two independent outputs in the unit, and there is an output

Figure 10.3
*Examples of
pneumatic control
systems (By
courtesy of Plessey
and of Maxam
Power Ltd.)*

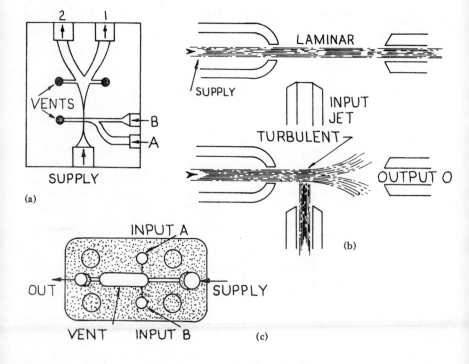

when neither input 1 NOR input 2 is present.

Other logic units are formed from this basic NOR unit, e.g. a NOT unit is similarly constructed but with one input. An AND unit connects the output of two standard NOR units to the input of a third NOR unit. An output is obtained when the inputs are both present, as in any NOR system.

Units from the Plessey Co. Ltd depend upon the disturbance of the air flow through a restriction in a plastics block (Figure 10.3(a)). Without inputs applied, the air supply will flow through the unit and out at output 1 only. When the module feeds three similar units, the output pressure will be 0.2 times the supply pressure, which may be between a minimum of 0.05 kg/cm^2 and a maximum of 1.7 kg/cm^2. With only one unit connected to the output, a higher pressure is obtained.

The application of an input at right angles to the air flow through the restricted path deflects the flow from output 1 to output 2; thus there is a switching action when an input is applied to either A or B. This module has an output at 1 when neither A nor B is applied, and it may be used as a NOR unit. There is also an output at 2 when either A or B is applied. It is therefore called an OR/NOR type of unit, with inverting and non-inverting outputs. The minimum control pressure required is about 0.1 times the supply pressure, so that the outputs may be at a greater pressure than the control pressure. There is an amplification of the control signals in the unit, so that a number of modules may be interconnected to form a complex switching system. The size of the unit shown is 5 cm x 4 cm x 1 cm.

A fluid logic element by Maxam Power Ltd, (Figure 10.3(b)) is based upon the principle of the turbulence amplifier. through the unit, between the supply and output, there is normally a laminar stream of air, while a very small input jet at right angles to the flow is sufficient to disturb the laminar flow and to produce turbulence when the air passes out through a vent. The supply pressure is about 0.02 kg/cm^2 with an air consumption of 0.08 m^3/h. Output pressure may be about one-half that of the supply, and control pressure one-eighth of this value, so that pressure is amplified between control and output.

The industrial type of turbulence amplifier may have four input tubes (Figure 10.3(c)), and is a basic NOR unit from which other logic elements can be built up. The planar turbulence amplifier consists of two plastic wafers, smaller than the area of a razor blade. The element is shown with two inputs, and units assembled in banks with integral supply connections form a very compact control system with a 'switching' time of 1.2 ms.

Landis Grinding Machines and Sizing Developments

The previous sections on automatic sizing on machines have shown that the means employed to ensure accuracy in work production are many and varied, but the first grinding machines to use programmed numerical feed to ensure uniform and reliable grinding cycles were of Landis design (Landis-Lund Ltd, Crosshills, Keighley). Apart from plain grinding machines, a range of high-production models is available, many of them being of special design for the automotive industry where mass production of components such as crank shafts, cam shafts and similar units must be produced rapidly and with consistent accuracy over large work batches.

A typical example from the Landis range is shown in Figure 10.4; this depicts a multiple-wheel grinding machine illustrating the grinding of crank shafts with all five wheels in operation together, along with the Marposs sizing heads. The machine has programmed numerical feed and solid-state control, and the elements of control are shown in Figure 10.5.

Figure 10.4.
A multi-wheel grinding machine for crank shafts (By courtesy of Landis-Lund Ltd.)

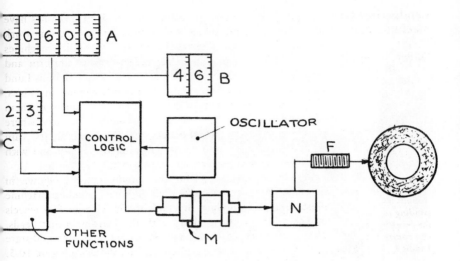

The control logic comprises the solid-state control for a numerical feed and machine cycle, while the numbered thumb-wheel switches (see Figure 10.6, A, B, C) can be set to pro-gramme the complete grinding feed and dressing cycles. Increased reliability is obtained by the integrated-circuit electronics shown in Figure 10.5. The digital commands from the solid-state programmed controller are accurately timed by a 400,000 c/s crystal-controlled oscillator accurate to within 1/10 of 1°. Thus any inherent grinding feed variables such as temperature change effects or variation of oil vis-cosity, often associated with hydraulic wheel feeds, are eliminated.

There is a co-ordinated function of digital control to a stepping motor M, which through a gear train N and a pre-cision feed screw F regulates the rate, distance and direction of the wheel head. It also stops the screw for required dwell periods.

The Landis Electronic Micro-feed automatically adapts the numerical grinding feed to compensate for wheel wear and machine temperature variations, which may affect work size, and allows a fine pulse feed in 0.00125 mm increments for approach to finish size. An illuminated visual read-out on the front control panel shows the progress of the wheel feed, while the cycle sequence is indicated by words which illumin-ate on the front control panel.

Stepping Motors

Reference has been made to a stepping motor, and there has been a marked increase of interest in these motors as a means of providing a precise discrete incremental drive to NC

Figure 10.6.
*The control logic
system for a Landis
machine:*
*A, B & C thumb-
wheel switches*

machine tools for slide-positioning duties. One of the major
advantages offered by this type of drive is that it is of the
'open-loop' category, i.e. it requires no separate positioning
feedback system which measures the actual distance travelled
by a moving member, compares it with the demanded distance
and uses the error signal resulting from differences between
these two values to initiate drive in the required direction.
Moreover, the stepping motor offers the opportunity to drive
the traverse screw directly, thus simplifying the transmission
and eliminating possible sources of error.

A stepping motor may be described as an actuator which
transforms an electrical impulse into a mechanical displacement of an incremental rotary nature, and it may be likened,
from a function point of view, to a solenoid-operated ratchet.
Different types of motor are available, including low-torque
forms providing frequencies up to 20,000 pulses/s; forms
having slow-speed torque ratings are also available, but the
three basic designs are characterized by the type of rotor,
which may be of permanent magnet, active or reactive type.
The latter is used most often for machine tool purposes.

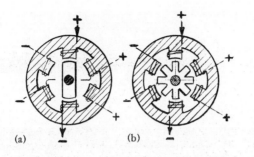

(a)   (b)

Figure 10.7.
*The operating
principle of a
stepping machine*

The operating principle, often referred to as the variable-reluctance principle, is shown in Figure 10.7(a). The motor is represented as a three-phase, six-pole stator and a two-pole rotor unit. The excitation of any single phase causes the rotor to be aligned by taking up the position at which the minimum magnetic reluctance is offered to the magnetic flux set-up. If the phases are now excited in true sequence, i.e. with current reversal where necessary to maintain polarity, the rotor will be indexed to provide rotational movement in increments of 60°. The value of an increment, also known as the stepping angle, is directly proportional to the number of stator poles, although it can be halved by exciting two adjacent poles at a time, thus establishing intermediate positions for the rotor.

In practice stepping angles of only a few degrees are required, and, in order to obtain these small angles, a multi-rotor element is employed. Figure 10.7(b) shows a motor with an eight-pole rotor and a six-pole stator, and, by excitation as before, this causes the rotor to index in increments of 15°. The general expression for calculating the step angle in degrees for a multi-pole multi-phase arrangement, based on single-phase excitation, is

$$\frac{360}{\text{number of poles x number of phases}}$$

Figure 10.8 shows the practical arrangement for a multi-pole reactive rotor unit. There is a series of co-axial units each acting on one phase of the multi-phase supply current. The angular positions taken up by the combined rotor as the elements are excited sequentially provides a corresponding series of increments totalling one complete revolution. Each element operates on one phase of the supply, so that the diagram represents a six-phase motor. One design of this basic type has sixteen poles in total for the rotor, and of similar design is the Fujitsu motor which is of the five phase, thirty pole type, but in this unit the rotor elements are in the form of discs with printed circuits.

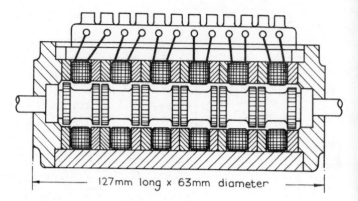

127mm long x 63mm diameter

Figure 10.8.
*A practical
arrangement for a
multi-pole reactive
rotor unit*

**Electro-
hydraulic
Stepping Motors**

For some machine tool or large inspection machine applications, a stepping motor may have the unfavourable characteristic of low acceleration and slow stepping speeds, and in some instances a combination of a stepping motor and a hydraulic torque amplifier may be used. Termed an electro-hydraulic stepping motor, Figure 10.9 shows the Fujitsu design in which the hydraulic section consists of a fixed-angle swash-plate axial-piston motor. The valve spool is connected at one end to the spindle of the motor by a screw and nut, and at the other to the stepping motor through gearing. When the stepping motor is operated under a control command, the spindle rotates the valve spool, causing it to be displaced axially under the action of the screw and nut. The hydraulic motor starts to operate immediately in such a direction as to 'wind' the spool back to the neutral position, and these conditions can be sustained, altered or terminated by suitable control of the stepping motor, thus causing a final output from the combined unit which is determined according to the electrical input command signal.

Figure 10.9.
*The Fujitsu
electro-hydraulic
stepping motor*

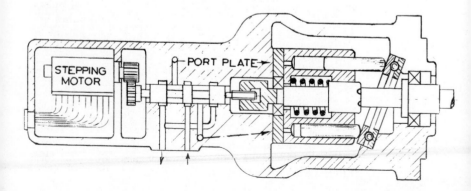

STEPPING
MOTOR   PORT PLATE

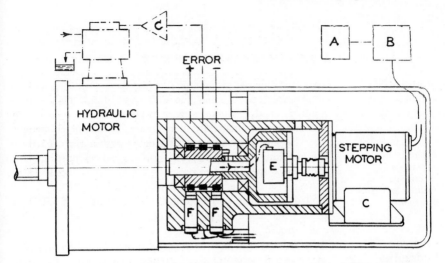

Figure 10.10.
*A stepping motor
for accurate
positioning of
machine slides:*

A   control unit
B   drive unit
C   servo-amplifier
E   potentiometer
F   brushes

Another design of electro-hydraulic stepping motor of particular interest for slide positioning is shown in Figure 10.10 (Dowty Hydraulic Units Ltd). The diagram shows the complete system, the input being derived from the control unit A and the drive unit B. Pulses are fed from the drive unit to the stepping motor; this, in turn, drives the wiper of a potentiometer E connected to the hydraulic motor and is driven by the latter. Basically, the system is a null-seeking circuit in that the body of the potentiometer is driven to follow the movements of the wiper blade. In fact, it is the error which exists between the relative positions of the wiper and the body which provides a signal that is picked up by brushes from slip rings.

The error is fed to a servo-amplifier C and thence to a hydraulic servo-valve which provides for movement of the hydraulic motor in the required direction, to follow the movements of the wiper blade. The shaft extension at the end of the hydraulic motor is connected to the traverse screw of the slide to be positioned. Thus, from the control unit, data are transmitted in terms of the number of pulses required to move the distance desired, and these pulses are injected into the stepping motor by way of the oscillator in the drive unit. A feature of the system is that the frequency of the pulses can be controlled and thus the speed of traverse of the slide or saddle. It can be arranged that for any distance to be travelled the greater part can be at high speed, and a 'creep' speed, resulting from automatic reduction in pulse frequency, can be engaged to ensure final accurate positioning in either direction.

Initially the standard increment of the motor is 100 steps/ rev, and the frequency of pulse transmission can range up to

12,000 pulses/s. The system is capable of positioning a slide to within 0.012 mm. In the arrangement of slip rings, two brushes F are provided for each ring.

Landis NC Grinder

This machine is programmed by a data-processing card to grind automatically multiple diameters at one loading. Any number of diameters on a workpiece can be programmed and ground automatically in sequence, the diameters being of any size and combinations of lengths within the capacity of the machine. All changes of grinding wheel position, feed rate, work speed and in-process gauging trip points are made in response to the punched card programme. Quality is controlled by direct and automatic measurement of the workpiece while it is being ground. Figure 10.11 shows the machine and console. The NCs are solid state with integrated circuits for reliability and optimum performance from the standpoint of electrical noise rejection, temperature stability and maintenance. The control console, seen on the right, includes digital read-outs to show the programmed diameter being ground. The card reader accommodates two data-processing

Figure 10.11.
The machine and console for a Landis-Lund NC grinder

cards to programme automatic grinding of shafts requiring as many as fourteen grinding stations. The rotational speed of the workpiece and the rate of stock removal are automatically set in relation to the diameter to be ground.

Automatic In-process Sizing

The electronic programmed sizing gauge is the major development which made NC for precision grinding practical. The gauge is seen mounted on the machine in Figure 10.12 and is responsive to 40,000 trip points through a 100 mm range of diameters as directed by the NC programme. It eliminates the need for an individual pre-set gauge for each different diameter to be ground and also serves as an adaptive control for the NC grinder. In conjunction with the wheel feed system, the wheel feed cycle is automatically adapted for each diameter to compensate for variables in wheel pressure, workpiece deflection, temperature changes and wheel condition.

After the first diameter is ground, the cycle is repeated for the next diameter. All new settings of the machine elements and the sizing gauge are made quickly and automatically after each diameter is ground, and all diameters are ground in the

Figure 10.12.
The Landis-Lund electronic programmed sizing gauge

programmed sequence. Normal size tolerances are held,
even when keyways and splines have been machined prior to
grinding. The sizing gauge automatically qualifies the workpiece for shoulder clearance, stock allowance and readiness
to grind. It will reject a workpiece either improperly prepared
for diameter and shoulder location or improperly programmed.

LANDIS SIZE-FINDER SIZING GAUGE

This instrument can cover an uninterrupted diameter range of
100 mm and can be used on components with up to eight
different diameters. It is a chordal shoe-type head, carbide-
lined, with a sealed linear transducer mounted between the
shoes to measure the workpiece diameter. The system operates from 110 V single-phase 50 c/s supply and has an accuracy
of repeatability of 2 μm at 20°C, and a range accuracy of
5 μm per 25 mm at 20°C, 7 μm per 50mm and 10 μm per
100 mm. There are three options available for the automatic
grinding cycle.

Hydraulic Feed

The operator sets the gauge channel setting for each diameter
to be ground. These settings include all feed control points
and on-size dimension signal. He then sets the wheel head
feed dial display indicator to the diameter dimension of that
surface of the component to be ground and activates the gauge
channel appropriate to that diameter. Thereafter the plunge-
grinding operation is automatic to the completion of grinding
of that diameter. The operator then re-sets the wheel head
feed dial to the next required diameter, activates the appropriate gauge channel, traverses the table carriage to align
the workpiece with the grinding wheel and initiates the automatic grinding cycle of this diameter. Subsequent diameters
are ground in like manner.

Microtronic feed

The gauge can be provided on Size-finder machines equipped
with this feed system (Figure 10.13). In this case the operator
merely dials the required feed by a small wheel feed positioner
knob and sets the gauge in precisely the same manner as
described before.

Multi-stop

Where repetitive batches justify the preparation of mechanical
programmes, the operator can be totally relieved of the
responsibility of setting the wheel feed position dial. This is
automatically achieved by coupling the wheel head in-feed
setting to the table carriage position, the gauge being still

Figure 10.13.
A microtronic feed system used on a Size-finder grinding machine

applied as described before. In all three cases there is no longer any need to stop and check with micrometers as a component is approaching finished size. Thus output can be considerably improved; to cite one case only, for lathe spindles the grinding time is reduced by 50%.

AUTOMATIC TAPER CORRECTION

Another feature of the Landis machines when grinding long workpieces, is that the system may incorporate two gauge heads (Figure 10.14), each feeding its own diameter-measuring amplifier, while a third amplifier compares the difference between the two diameter-measuring amplifiers and controls the operation of the taper compensator. The system is of particular value when grinding long shafts where slight taper may be difficult to detect, but with a gauge head at each end of the shaft a size meter can give a visual indication of the size at each end, while a centre meter shows the amount and direction of the taper. The taper-operating mechanism is in contact with each of the wheel bearings on the spindle and by hydro-mechanical means corrects any out-of-tolerance taper.

The two sizing meters are graduated in $\frac{1}{100}$ mm, while the centre zero meter, indicating the difference between the two diameters, is also of the same scale graduated – 30–0– +30, plus and minus signs indicating taper direction.

Figure 10.14.
Gauge heads fitted on a Landis-Lund grinder for automatic taper correction

11.
The Inspection Department

Introduction

All consumers want quality in the widest sense of the term, i.e. fitness for purpose. Good design and standards in themselves must lead towards quality, but a lowering of the standards of quality may result from manufacturing errors and from the human factor in careless work, so that this feature brings in the need for inspection. Inspection costs money, and one problem of management is to provide a degree of inspection that will ensure adequate quality and reliability without unduly adding to the cost by an inspection system which is too elaborate and with standards unnecessarily precise.

Organization

In the organization of inspection departments, substantial independence of control is a prime essential, while correlation with other departments must be maintained. Should the chief inspector be subservient to an official whose main preoccupation is output, obvious difficulties may arise. A sound arrangement is that the chief inspector should be responsible to the general manager who controls all aspects of the organization (productive, technical and commercial) and who is therefore likely to have an objective approach to any problems which may arise regarding rejected work.

Inspection should be concerned with everything listed entering and leaving the factory, i.e. (1) raw materials, (2) components bought out, (3) manufacturing operations, (4) sub-assemblies, (5) final erection and (6) tools and gauges. The majority of inspection staff are likely to be employed in the manufacturing branch. This is especially so in large-scale production such as the production of motor cars where batteries of automatic machines are used and where hold-ups on the assembly line cannot be tolerated. The ideal inspection procedure for machining operations is that each part is stamped after inspection at every operation so that no further work is done on faulty components. This method, however, may be costly and may cause delay. No hard and fast rules can be laid down, because inspection procedure must be planned to suit the type of factory, the class and function of the product and the economics of the situation.

Line Inspection

This is a system in which the inspector visits the machine and checks the work while machining is in progress. This method is common in large factories and has the advantage that faults can be detected before much bad work has been produced and also that rejects can often be corrected in the machine before the 'set-up' is broken down. Against this, too much inspection on the shop floor tends to create the impression in the mind of the operator that the maintenance of quality in production is none of his business, a state of affairs to be avoided at all costs. Whilst an inspector must bear responsibility for the work he approves, the production department and the operator or machine setter must be held equally responsible if good standards and low scrap percentages are to be realized.

The importance of materials inspection rises with an increase in the use of automatic machines of high output, because serious losses from poor material can then arise so quickly. It is probably true to say that 100% inspection of work is a practical impossibility, and certainly non-productive labour has to be kept to a minimum in a highly competive industry. The value to a firm therefore of an individual inspector (or an inspection department) largely depends on the inspector's ability to select for inspection those details of a production operation or component which foresight and experience indicate to be the most critical.

The standard of work produced in the machine shop is closely dependent on the accuracy and suitability of the tools, jigs, fixtures and gauges used. Tool room inspection is therefore most important in that the inspection equipment, length standards, comparators and measuring instruments must be of a higher degree of precision that the machine shop inspection equipment, owing to the fact that the manufacturing tolerances on gauges can only be a fraction (of the order of 10% normally) of the tolerances provided on the component. Thus it is apparent that instruments such as hand micrometers are not suitable for such fine measurements, and in modern tool room or inspection departments a wide range of precision instruments are employed, some capable of detecting errors as small as a few millionths of an inch or small parts of a micrometre.

Reducing Manual Inspection

The machines used for full automatic inspection have been described earlier, and the comments here refer to the reasons for the growth of automatic inspection. When automatic manufacturing methods are adopted in an engineering plant, the needs for inspection change. Increased reliance may be placed on the machines because they are not subject to the errors and fatigue of the operators, thus eliminating some of the inspection operations. Conversely the absence of an operator raises new problems, e.g. in an automatic transfer machine it may be necessary to provide a device which will detect a broken drill remaining in a hole and then will stop the machine in order to prevent the damage that would occur if an attempt was made to tap the hole. An inspection operation of this kind would unconsciously be performed by the operator. The comment is of importance, for on one transfer machine alone (Automotive Products Ltd) an electronic device detects some twenty broken drills per day.

Another example of special needs if an inspector is replaced concerns automatic assembly machines where a current problem is that of machine jams owing to components being outside tolerance. The solution again may well lie in the use of automatic inspection to eliminate incorrect parts, and similarly automatic inspection may be used to control processes. Here the inspection device is the means of measuring the controlled quantity.

The principal factors which influence the decision to use automatic inspection equipment may be summarized thus. The capital cost of the equipment is always greater than that for manual methods. When the manual method depends largely on the inspector and the use of simple equipment, the capital cost of an automatic instrument will be much higher. However, where the measurement is already made by an instrument and where the part played by the inspector is mainly to read and adjust it, the additional cost of having an automatic instrument and alarm signal when the tolerance is exceeded may be comparatively small.

Inspection on Transfer Line
Automatic inspection of diameters being ground on production grinding machines is well established, and developments in this field have been described earlier. They provide a good example of the way that production personnel are responsible during manufacture for the quality of the finished work. A similar technique is now being employed on transfer machines producing turned components.

Measurements are made on the transfer machine at carefully positioned stations, and the work is transferred to the inspection equipment by the same mechanism that transfers the work between the working stations. As tool wear becomes apparent, dimensional changes take place in the finished parts, and signals are sent from the inspection units to pulse motors which are coupled to the positive stops on the turning

Figure 11.1.
*Etamic inspection
probes in a transfer
line for brake discs
(By courtesy of Filli
Morando, Turin)*

machines. The stops are thereby automatically adjusted to
keep the dimensions within the required tolerance band, and
this is all done without stopping the machine. In normal cir-
cumstances the tools or tips would be discarded when an
extreme of tolerance limit was reached, and with in-process
inspection the tool life is also increased.

Figure 11.1 shows Etamic inspection probes in the transfer
line for Fiat 124 brake discs. A visual display of the discrep-
ancy from nominal size is given for each of two components
by the two dials situated on the heads, and three counters are

Figure 11.2.
A transfer line for machining and inspection of brake discs

provided for each inspection station to give statistical information on the production, tolerance, oversize, undersize being counted. The equipment also tracks a part which is in error and and diverts it from the transfer line at the earliest convenient point.

The inspection equipment installed in this line, made by Flli Morando, Turin, fully inspects the total production of the line which includes drilling, tapping, reaming and turning, passing four brake discs per minute. It controls the critical dimensions by altering the machine stops, and it also counts good and reject parts, sorting them before the parts go on to subsequent treatments. Figure 11.2 shows the complete line with three turning stations and the inspection station for the turned features which is followed by drilling, then another inspection station and finally for the reaming and tapping operations which are done at the end of the line.

Speed and Consistency

The time required to perform an inspection operation is important, because it governs the rate at which components can be inspected and so determines the number of instruments required for a given production rate. If measurement forms part of a control loop, it may determine the speed at which the process can operate and thus may fix the output of a large amount of equipment. In general, automatic instruments operate more quickly than manual methods owing to the speed with which deviations from the standard are detected and signals given, for the high operating speed of electronic circuits can be utilized. The gauging of small components to check that the dimensions are within the specified tolerance is the most intensive application, particularly in industries where precision components are made at high production rates, e.g. the automobile industry. There are two other considerations involved in automatic inspection operations. These are (1) the mechanical handling required to present the work to the measuring instrument and (2) the processing and recording of the test results.

Measurement

Measuring dimensions involves determining the distance between specified points on certain surfaces. For automatic measurement the basic problem is that of converting linear displacement into an electrical or other form suitable for transmission away from the inspection point. An exception to this is found in some 'go' and 'no-go' gauges which automatically divert the parts into different chutes depending on their size. Here the measurement and comparison functions are combined. Mechanical means for this purpose are limited to components of a suitable shape, and each system must be designed for its special purpose.

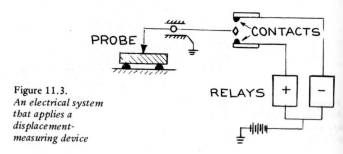

Figure 11.3.
*An electrical system
that applies a
displacement-
measuring device*

In electrical systems the usual way of applying a displacement-measuring device to inspection is shown in Figure 11.3. The workpiece is held with one of the defining surfaces registered against a fixed stop, a transducer being used to determine the distance between the fixed register and tip of the probe. The simplest type of transducer used for this work consists of a sensitive limit switch shown in the same diagram. The adjustable contacts are so arranged that, when the dimension in question is within tolerance, neither is closed. If the upper or lower tolerance is exceeded, one of the contacts is made, and current flows to energize a relay. This operates a selector mechanism in the mechanical handling section, causing the rejected part to be diverted into an appropriate container.

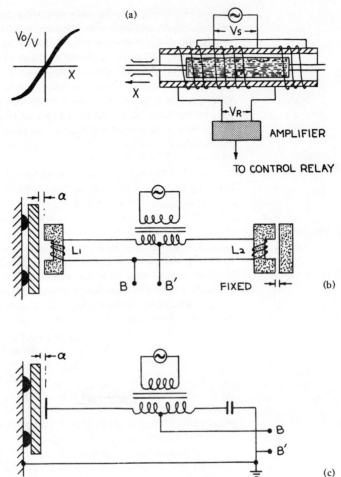

Figure 11.4.
Methods employing the air gap principle of inspection

Switches are available in a range of sensitivities, some detecting movement as small as 0.001 mm.

This type of detector forms a convenient basis for building gauging stations in an inspection department for checking a number of dimensions on a single component. They are simple and do not need electronic amplifiers to couple them to control equipment but have two inherent limitations: (1) they cannot give a continuously variable signal proportional to the dimensions gauged and (2) they cannot be used to grade parts into more than three categories. For this and other duties such as process gauging, electromagnetic transducers are used, as they are sensitive and robust. The movement of a probe causes either a change in the reluctance of a magnetic circuit or a change in the coupling between two coils; Figure 11.4(a) is an example of a linear differential transformer.

The methods described have involved mechanical contact between the detector and the surface to be gauged, but methods are available which depend upon measuring the air gap between the surface to be inspected and a fixed probe. The principle is shown in Figure 11.4(b) where a change in the air gap dimension a changes the inductance L_1 of the coil so that the bridge unbalance voltage at the terminals B, B$'$ is a direct indication of the gap a and hence of the dimension required. This method indicates the principle of the electro-magnetic gauge, and a similar method in which the electrostatic capacitance between the work and a fixed probe is used is shown in Figure 11.4(c).

Computerization It is well understood that an inspection department has a primary function of ensuring that a laid-down standard of quality is maintained, and this feature is accomplished in various ways. It must ensure that raw materials conform to the standard required and that subsequent work put into those raw materials is again up to the correct standard; this should embrace packing and shipping too. However, it is not always realized that the process of inspection serves as a monitor with a data feedback on the performance of the manufacturing and processing departments. It can supply much useful information; in the case of mass production industries it can also predict tool life on a new job, and frequently a line inspector can make a real contribution towards lowering manufacturing costs. The data which have to be handled often require a great number of man hours to process, and, just as computers are being put to work to plan production schedules and to pay factory wages, they are now also being employed in the processing of data pertinent to quality and the functions of the inspection department.

Thus consideration is now being given to setting up computer-controlled 'inspection centres', in some ways similar to machining centres already in use in the manufacturing division and so called because they are able to tackle a wide variety of jobs with very little time wasted between operations. The investigation in a factory making aero engines concerns a computer to programme NC machine tools and at the same time to programme the inspection centre to accept or reject the part immediately it has left the machine.

The inspection centre embodies various stations at which some type of measurement is made. There are stations for linear, rotary and diameter measurement and a central computer to insert a programme into the inspection centre when the part leaves the machine and when the centre accepts the

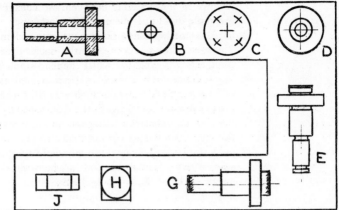

Figure 11.5.
*An inspection
centre with stations
for checking
components:*
A *outside diameter
   check*
B *bore check*
C *hole centre check*
D *concentricity
   check*
E *grooves and
   undercuts*
G *length checks*
H *hardness test*
J *stamp O.K.*

part. Each station is programmed to check one or a series of dimensions in the correct order, as shown in Figure 11.5, and to compare results with a pre-punched tape containing all the relevant information on that part. The inspector has a route card to follow around the inspection centre giving the various checks which have to be made at the various stations shown in the figure. Inspection on the line could be carried out in this way by the machine setter or operator, who in the case of a production shop may have to wait for clearance on his first part before proceeding with the remainder of the batch.

Thus the inspection department is fully integrated into the factory system; the computer is arranged to process orders from the customer by ordering from the manufacturing unit the parts needed, giving the purchasing department a print-out of parts which have to be bought out and printing an acknowledgement of the order to send to the customer, with delivery date. The computer follows this up with a critical path analysis of the manufacturing schedule with due regard to work already in hand and produces lists of tooling required for the various parts and tapes for the manufacturing and inspection centres. Feedback information from the line to the computer is essential so that performance can be monitered, and a nightly print-out summarizing performance against planned activities should be given. Additionally, feedback information is given to enable the next programme to be ready for each machine or inspection centre when it is required. It will be seen that under ideal conditions neither the operator of the machine tool nor the inspector needs a drawing of the component being handled.

The system described is being used in a number of aircraft factories in both Europe and America. Design information on the whole of the aircraft is stored on tape and can be called by the computer. For instance, if a drawing is needed to give an idea of basic shape to the operator, an outline drawing of, say a wing section can be made in a matter of minutes by a tape from the computer used in a drawing machine.

The computer is used to produce the manufacturing tape and to give a print-out of 'geometric data' which is used by the inspector as his drawing. Often, the inspector may not need all the information given to the machinist as on continuous-path milling, for instance, several point checks on a profile are usually sufficient to establish whether or not the part is acceptable, whereas, for machining, continuous-path information is needed. The inspector may therefore make a short punched tape for his inspection machine. Indeed, at the present time, expertise is needed in programming a computer to eliminate non-essential information for the inspector, particularly information given for machining which is subsequently cut away in a later process.

Inspection of Aircraft and Automobile Components

The change-over from manually operated and controlled machine tools to those actuated by magnetic or punched tape has brought problems into many inspection departments, the most serious being that of time, e.g. a single aircraft part formerly produced by a normal routing machine took 6 h to produce, while with a NC machine three parts are now produced simultaneously in 2 h. The result is that an inspector can be inundated with work in a very short time. As an example let us consider the results obtained by old and new methods on the given component and shown in Table 11.1.

Table 11.1

	Old method	New method
Production time, one off (h)	6	0.6
Measuring time, one off (h)	0.5	0.5
Measuring/production (%)	8.3	83

Thus, the inspection time must be reduced by at least 75% to keep up with the pre-NC time ratio.

It should be remembered that modern assembly methods have reduced times in production, so that machined components are required by the assembly shops at more frequent intervals of time than previously. Unless extra inspection staff and more measuring machines are installed in the works, and this is to be avoided if possible, the inspection department will be overloaded. It therefore becomes necessary to reduce inspection times by more than 75%, and means will be indicated to show how inspection time can be reduced by the utilization

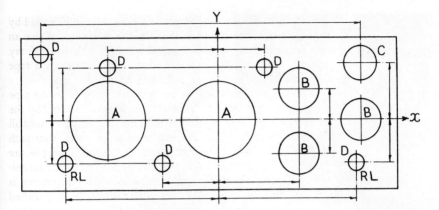

Figure 11.6.
*The example of
checking bores in
one plane using
conical probes:*

*A, B, C & D hole
centres*
RL reference holes

of modern inspection machines.

When an inspection machine is installed, a consideration of the equipment required must be made. The simplicity or complexity of this feature depends upon the type of work to be inspected, and in Chapter 5 the types of probes available have been indicated. If, as shown in Figure 11.6, it is simply a matter of checking bores in one plane, then simple cone probes of suitable diameters will suffice. Some co-operation with the drawing office is required so that the inspector can work from datum points without having to make any calculations. The example given is that of checking bored holes on an Olivetti machine; the procedure is as follows.

(1) The workpiece must be positioned and aligned on the reference holes RL by an angular table adjustment.
(2) It must be centred on the origin of axes with the cone probe; X and Y must be zeroed and the dimensions of the hole centres A must be read.
(3) With the smaller cone probe the dimensions of the hole centres B, C must be read.
(4) With the smallest cone probe the dimensions of the hole centre D must be read.

The time required for positioning and aligning the workpiece is 4 min, and the time required for checking with the cone probes is 3.5 min; in total this is 7.5 min. With conventional measurement the time required for twice positioning and alignment is 9 min. The time required for checking is 16 min, and therefore the total time is 25 min.

AIRCRAFT COMPONENTS
It is in this industry that large and complicated components appear for inspection, so that comparatively large and well-

Figure 11.7.
*Inspection using a
surface table and
vernier height gauge*

equipped machines are required. Speed of operation is required, but accuracy is also essential. The tolerance on aircraft components is usually demanded to be around ±0.25 mm, but sometimes the accuracy required may reach ±0.07 mm.

In one prominent British aircraft factory, two inspection machines are installed, one a Ferranti and the other, used for the larger components, has a capacity of 2,800 mm × 1,600 mm × 1,000 mm in the X, Y and Z axes respectively. The measuring probes are inserted in the lower end of the square ram in any of five positions, i.e. probes may point left, right, forwards, backwards or downwards.

The attachment of the measuring probes on the head is of the electro-magnetic type in which a permanent magnet is incorporated in the base of the probe, while in every face of the machine head is fitted a magnet keeper. To eliminate the magnetic action at the time of the application or removal of the instrument, an electro-magnetic field with opposite polarity is generated by inserting a rapid connecting plug. This causes no undesirable displacement at the moment of locking. The reference basis for the application of the instruments is the horizontal shoulder and sides of the head parallel to the co-ordinate planes.

A comparison of the old method of inspection using a surface table and vernier height gauge is shown in Figure 11.7 (a), (b); the work is checking co-ordinators, and the new method of checking using the measuring machine is shown in Figure 11.8.

Figure 11.8. opposite
*A replacement
method using
measuring machine
on aircraft
components (By
courtesy of Hawker-
Siddeley Ltd.)*

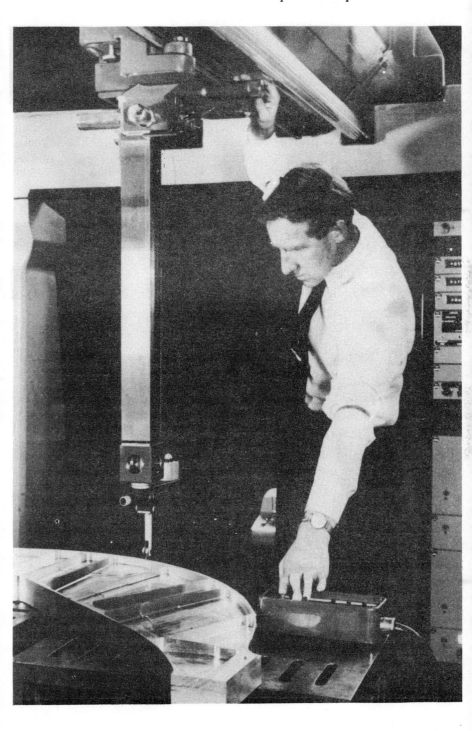

EXAMPLES OF PARTS CHECKED
The operations performed and the time comparisons for conventional means and by the measuring machine are listed.

Aileron hinge rib
1,016 mm × 178 mm × 38 mm; five off; contoured and varying angle faces on two long sides; machined pockets on one large face and two holes 38 mm in diameter in web; twenty-eight co-ordinate points on contoured faces, stiffener and hole size and positions, and flange thickness checked to programme compiled from loft template; time taken, 2½ h; previous time, 12 h; time saved, 9½ h = 80%.

Rear wing spar
1,400 mm × 127 mm × 44 mm; sixteen off; contoured and varying angled faces on two long sides; machined pocket on one large face and holes in web; twenty-six co-ordinate points on contoured sides checked to dimensions on drawing; time taken, 5¾ h; previous time, 30 h; time saved, 24¼ h = 82%.

Top fitting
1,524 mm × 305 mm × 25 mm; five off; contoured faces on four sides; machined pockets on one face; forty-six co-ordinated points checked to a programme compiled from loft template; time taken, 3 h; previous time, 10 h; time saved, 7 h = 70%

Nose rib
660 mm × 304 mm × 50 mm; five off; contoured and varying angled faces on two long sides; machined pockets and apertures on one large face; fifty-three co-ordinate points on contoured faces, stiffeners and apertures, checked to a programme previously compiled; time taken, 2½ h; previous time, 12 h; time saved 9½ h = 79.2%.

CHECKING OF TRACTOR HOUSING CENTRE
The unit is shown in Figure 11.9 with the points of inspection indicated by the numbers. The tools used for the operation are (1) a swivelling surface gauge with $^1/_{100}$ mm dial gauge (2) a square block reference gauge (3) a magnetic base and (4) an electronic touch finger for hole centre distances and shoulder measurements. The overall dimensions of the workpiece are 860 mm × 445 mm × 275 mm with tolerances required up to 0.025 mm.

The workpiece is positioned on the surface plate on three adjustable supports indicated at S (two) and T. The housing centre is thus placed in such a way that the higher surface E is facing the tool-carrying ram, and its centre line is parallel

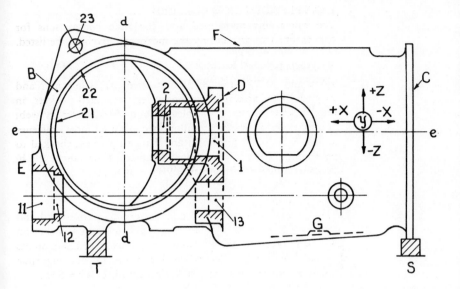

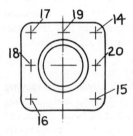

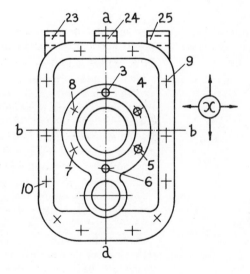

Figure 11.9.
*An example of
checking dimensions
on a tractor housing*

with the longitudinal machine axis X. This position allows the execution of all the inspection operations and gives free access to the important area of the workpiece without the need to turn over and re-set. Table 11.2 gives a summary of the inspection operations.

Table 11.2

Operation	Machine axis	Time
(1) Position workpiece		
(2) Align and level workpiece	−Z	9 min
(3) Position and align magnetic base and square reference gauge	−Z	2 min 30 s
(4) Determine axes, bores 1, 2 and check position of holes 3, 4, 5, 6, 7, 8, 9, 10; check surface G	+X	6 min 30 s
(5) Check centre distance of bores 11, 12, 13 and check position of holes 14, 15, 16, 17, 18, 19, 20	−X	5 min 30 s
(6) Determine axes and check concentricity of bores 21 and 22; check position of holes 23, 24, 25 and face E	−Y	4 min 30 s
(7) Check faces A, B, C, D and verify parallelism and squareness	−Z	5 min
		Total time 33 min

Control Gauge for Silencer Tube

An example of the way that three-dimensional measuring machines can help the inspector and the production department at the same time can be found in the various industries which manufacture precision pipe work. The aircraft, motor car and truck industries have closely allied problems where reasonably accurate pipes must be made to fit closely to the contour of a large body, whilst contact must be avoided with the moving parts that these vehicles have on their underside. A brief résumé of current methods is needed to understand the problems involved and to see how accuracy can be improved and production costs can be reduced.

Many motor car manufacturers, for instance, make prototype pipes for exhaust systems by hand bending, welding, and when a satisfactory shape is arrived at, they proceed to try to duplicate this in quantity. The first basic essential is to have either a set of dimensions which can be reproduced or a master which can be used for comparative inspection. Any part of a pipe which has bends requires three axis co-ordinates for a point plus two pieces of angular information showing which direction that part of the pipe is pointing relative to a datum line and its perpendiculars. As the pipe may have several bends, the amount of dimensional and angular information needed is considerable, and there are severe problems in holding samples of the pipe, together with practical difficulties in taking measurements.

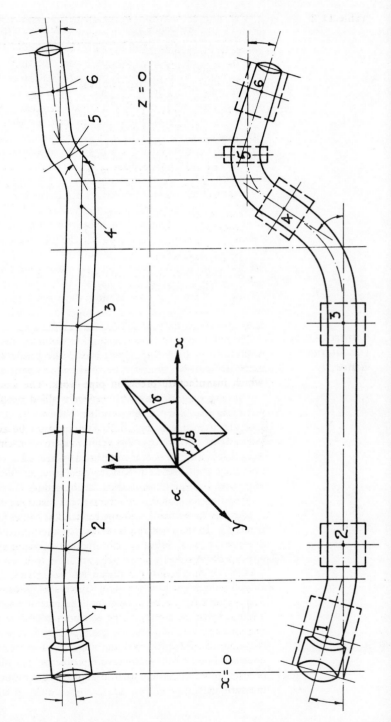

Figure 11.10.
*A method of
checking Fiat
exhaust pipes*

The normal method is therefore to make a wooden or plastic acceptance jig from the original hand-made pipe which can then be used as a sub-master. The acceptance fixture has accommodation points at critical parts of the pipe, usually on the bends, and no dimensional information is used except for the overall length of the pipe. If pipes are being manufactured incorrectly, it is purely a matter of guesswork to apply tolerances, and correctional information to the production pipe bender is difficult to define. The acceptance fixture is very bulky, particularly for a large car, and should always be near to the production unit as it takes up a lot of valuable space.

Because the acceptance fixtures are bulky it can be appreciated that they cannot be transported readily from store to inspection department and that therefore they are likely to be stored near to inspection personnel. The pipe benders requiring a check on their work must then carry a bulky workpiece to the fixture for the frequent quality checks which they must make because overbend varies a little with pipe material variations. In a large production shop with several current models of car being catered for together and with spares for many previous models it can be appreciated that a large conglomeration of these acceptance fixtures exists, two of each really being required (but rarely being provided because of the difficulty of accurate duplication) to satisfy the requirements of both production and inspection.

The example of the application of a three-dimensional measuring machine to this problem shows how a fixture for final acceptance can be eliminated using a system of loose fixture parts which can be assembled quickly to make an acceptance fixture. Adjustable saddles are used to hold the pipe; they can be set at almost any angle and can be positioned either on a base plate or on the measuring table (Figure 11.10).

A mating part used to position them is mounted in the ram of the measuring machine (Figure 11.11). Thus an inspection fixture for these pipes can be built up quickly when required and can be dismantled or changed to a fixture for another pipe equally rapidly.

In the example shown, a plan and front view of the pipe are given together with the most convenient positions for the locations saddles to seat the pipe. The centre position of the saddle is specified by the three cartesian co-ordinates X, Y, Z and the angle of inclination in plan and front view of the axis of the saddle. Having a fixed datum on the measuring machine the setting piece is inserted in the ram of the machine with its barrel inclined at the compound angle shown in the figure for that part of the pipe. The ram is then moved to the position

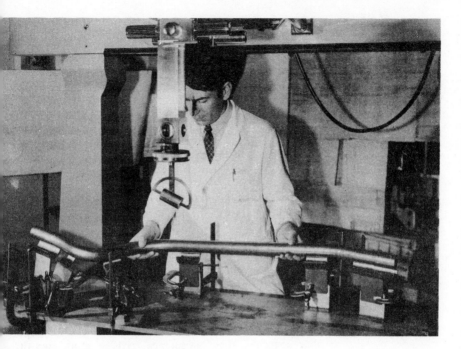

Figure 11.11.
*A measuring
machine and
inspection fixture
for exhaust pipes*

of the three co-ordinates, a saddle is positioned in contact with it and its clamps are fixed.

The process is repeated for the remaining saddles, and an acceptance fixture is ready for the pipe. Table 11.3 shows the three co-ordinates and the two angles that are needed to set each saddle.

Table 11.3

Point	β	α	X	Y	Z
1	1°	14°	67	28	166
2	2° 30′	0°	241	0	165
3	2° 30′	0°	684	0	146
4	0°	56°	913	129	140
5	26°	0°	1,032	204	173
6	6°	15°	1,135	183	135

The time taken for making the fixture is analysed as follows: time taken for alignment of setting pieces in ram, 3 min; time taken for positioning of six saddles, 18 min; total time taken, 21 min. The pipe can then be visually inspected. If a permanent fixture is required, the saddles can be electrically welded to the base plate, and each of the joints can also be welded to fix them in the correct position permanently.

12.
Worked Examples in
Inspection & Measurement

Problems in metrology are of a widely varying character, and, while by means of the attachments available on modern inspection machines practically all sizing problems can be solved, a great deal of inspection can be performed without resort to expensive and elaborate equipment. Most of the examples listed involve the use of trigonometry, and some practical experience is usually necessary to analyse a practical problem before attempting to apply mathematical calculations. The problems presented are of a practical nature including methods frequently applied in tool room and factory inspection departments. They can also be used for projects in the metrology room of educational establishments. The essential

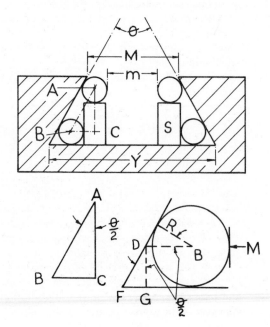

Figure 12.1.

equipment required comprises a set of slip gauges, two sets of precision rollers, two similar sets of balls, a surface plate, a sensitive indicator and straight edges.

Balls, Rollers and Slip Blocks

EXAMPLE 1

Figure 12.1 shows a V-shaped slide in which the sides of the dovetail must be made to match the mating surfaces sliding therein. It is usual to specify a dimension between the two points of intersection of the sloping sides and the base, while the angle of the sides must be accurate. Two rollers of equal diameter are placed in opposite corners of the dovetail, and a slip pile is inserted between them. The two cylinders are then vertically raised, and a second measurement of the gap between them is taken by slip gauges. From these dimensions the intersecting dimension Y and the angles of the flanks can be determined.

Let the radius of the rollers be R. Then

$$\tan\left(\frac{\theta}{2}\right) = \frac{BC}{AC}$$

$$BC = \frac{M - m}{2}$$

$$AC = S$$

and $$\tan\left(\frac{\theta}{2}\right) = \frac{M - m}{2S}$$

Also $$\sec\left(\frac{\theta}{2}\right) = \frac{DB}{R}$$

$$DB = R\sec\left(\frac{\theta}{2}\right)$$

$$\tan\left(\frac{\theta}{2}\right) = \frac{FG}{DG} \text{ or } \frac{FG}{R}$$

$$FG = R\tan\left(\frac{\theta}{2}\right)$$

Now $$Y = M + 2\left[R + DB + FG\right]$$

and $$Y = M + 2\left[R + R\sec\left(\frac{\theta}{2}\right) + R\tan\left(\frac{\theta}{2}\right)\right]$$

or $$Y = M + 2R\left[1 + \sec\left(\frac{\theta}{2}\right) + \tan\left(\frac{\theta}{2}\right)\right]$$

In measuring the angle θ, the vertical distance through which the cylinders are raised should be as large as possible. This requires a small roller, so that contact between the rollers and the flanks of the dovetail is made in a plane near to the base. A slip pile should then be used which allows the rollers to contact the dovetail sides in a position close to the mouth. This is because any error in measurement of this angle is mainly due to inaccuracies in the measurement of the two rollers M and m, and the angular effect of such errors is reduced as the distance between these two planes is increased.

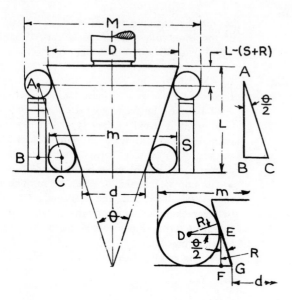

Figure 12.2.

EXAMPLE 2

To obtain the included angle of taper plugs (Figure 12.2), rollers are again used; the measurement is taken over the rollers by micrometer, or, if greater accuracy is required, the micrometer should be used as a comparator with slip gauges as the standard of reference. The uniformity of taper may be determined by making a third diameter check mid-way between the others and by comparing this with the theoretical dimension found by simple proportion according to the vertical displacements. For example, if M is 322 mm and m is 302, then the third dimension should be

$$302 + \frac{(322 - 302)}{2} = 312$$

It is also necessary to check the diameter of the taper in a given plane. This is normally specified in either of two extreme positions, i.e. the major and minor diameters, D and d. The method of calculation of the angle and both major and minor diameters is now given. Let the radius of the rollers be R.

$$\tan\left(\frac{\theta}{2}\right) = \frac{BC}{AB} \frac{M - m}{2/S}$$

or $$\tan\left(\frac{\theta}{2}\right) = \frac{M - m}{2S}$$

and the included angle θ is twice the angle $\frac{\theta}{2}$. The minor diameter d is given by

$$d = m - 2(R + DE + FG)$$

Also   DE $= R \sec \left(\frac{\theta}{2}\right)$

and   FG $= R \tan \left(\frac{\theta}{2}\right)$

Thus $d = m - 2\left[R + R \sec \left(\frac{\theta}{2}\right) + R \tan \left(\frac{\theta}{2}\right)\right]$

or $d = m - 2R\left[1 + \sec \left(\frac{\theta}{2}\right) + \tan \left(\frac{\theta}{2}\right)\right]$

In the same way, the major diameter D is given by

$$D = M - 2\left[R + R \sec \left(\frac{\theta}{2}\right)\right] + 2\left[L - (S + R)\right] \tan \left(\frac{\theta}{2}\right)$$

or $D = M - 2R\left[1 + \sec \left(\frac{\theta}{2}\right)\right] + 2(L - S - R) \tan \left(\frac{\theta}{2}\right)$

EXAMPLE 3

For checking internal tapers, balls must be used instead of rollers. The small diameter of the taper must be placed uppermost; otherwise the balls tend to be forced upwards when the slip blocks are inserted. As shown in Figure 12.3, it is sometimes necessary to raise the component from the surface table in order to increase the distance between the measuring planes. The derivation of the formulae for the included angle θ and the diameters D and d now follow.

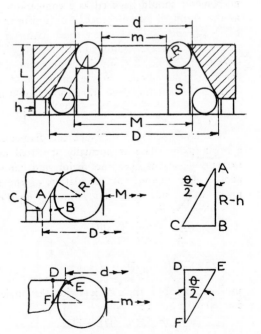

Figure 12.3.

$$\tan\left(\frac{\theta}{2}\right) = \frac{M-m}{2/S} \text{ or } \tan\left(\frac{\theta}{2}\right) - \frac{M-m}{2S}$$

$$D = M + 2\left[R + R\sec\left(\frac{\theta}{2}\right) + (R-b)\tan\left(\frac{\theta}{2}\right)\right]$$

$$\text{or } D = M + 2R\left[1 + \sec\left(\frac{\theta}{2}\right)\right] + 2(R-b)\tan\left(\frac{\theta}{2}\right)$$

$$d = m + 2R + 2R\sec\left(\frac{\theta}{2}\right) - 2DF\tan\left(\frac{\theta}{2}\right)$$

$$\text{or } d = m + 2\left[R + R\sec\left(\frac{\theta}{2}\right)\right] - 2\left[(L+b) - (S+R)\right]\tan\left(\frac{\theta}{2}\right)$$

EXAMPLE 4

To check the included angle and the extreme taper diameters of small taper bores (Figure 12.4), the larger ball should contact the sides of the taper near to the mouth with a smaller ball positioned near the small end. To measure the height over the balls, a vertical comparator will give very accurate results; otherwise two equal slip piles can be placed at each side of the work, and a depth micrometer can be used to measure the distance down to the top of each ball. As before, formulae can be calculated thus:

$$\sin\left(\frac{\theta}{2}\right) = \frac{R-r}{X}$$

and
$$X = (H-R) - (b-r)$$
$$d = 2(AB - DE)$$

Now
$$\sec\left(\frac{\theta}{2}\right) = \frac{AB}{r}$$

and
$$AB = r\sec\left(\frac{\theta}{2}\right)$$

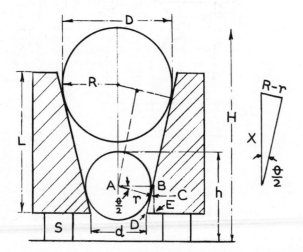

Figure 12.4.

Also $\tan\left(\dfrac{\theta}{2}\right) = \dfrac{DE}{BE}$

and $BE = b - (r + s)$

Thus $DE = [b - (r + s)]\tan\left(\dfrac{\theta}{2}\right)$

$$d = 2\left[r\sec\left(\dfrac{\theta}{2}\right) - [b - (r + s)]\tan\left(\dfrac{\theta}{2}\right)\right]$$

$$D = d + 2\tan\left(\dfrac{\theta}{2}\right)XL$$

EXAMPLE 5: MEASUREMENT OF PLAIN BORES

There are two methods using precision balls. In the first two balls are used (Figure 12.5(a)) and in the second case four balls, three of these of the same size (Figure 12.5(b)). The derivation of formulae for the determination of the bore diameter, given the ball dimensions and the measured height over the top ball for each example, are as follows. First for the two-ball method. The bore diameter D is given by

$$D = R_1 + CB + R_2$$

From Pythagoras' theorem

$$CB = [(R_1 + R_2)^2 - AB^2]^{1/2}$$

Now $AB = H - (R_2 - R_1)$

Therefore

$$CB = [(R_1 + R_2)^2 - [H - (R_2 + R_1)]^2]^{1/2}$$

Hence D can be evaluated if the diameters of the two balls are known.

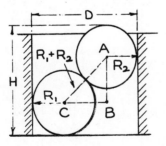
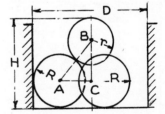
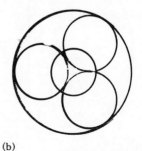

(a)

Figure 12.5.   (b)

For the four-ball method, the bore diameter is given by

$$D = 2(R + AC)$$

and $\quad AC = [(R + r)^2 - BC^2]^{1/2}$

Now $\quad BC = H - (R + r)$

Thus $\quad AC = [(R + r)^2 - [H - (R + r)]^2]^{1/2}$

Hence D can be found if the ball sizes are known.

One drawback to these methods is that the balls tend to be unstable in position and that it is sometimes difficult to obtain measurements over the upper ball. The difficulty can be overcome by the method shown in Figure 12.6, in which three balls are used.

EXAMPLE 6

The two lower balls are of equal size, and, as the balls are not free to move, the overall height dimension can be determined more easily. A general formula for the bore diameter using symbols can be derived as shown. If the lower ball has a diameter d_1, and the upper ball has a diameter d_2 and if the overall height over the upper ball is H, then

$$\text{diameter of bore} = \frac{A^2}{A - d_1/2}$$

Figure 12.6. $\qquad$ where $\qquad A = \left[\frac{d_1 d_2}{2} + \frac{d_2{}^2}{4} - \left(H - \frac{d_1 + d_2}{2}\right)^2\right]^{1/2} + \frac{d_2}{2}$

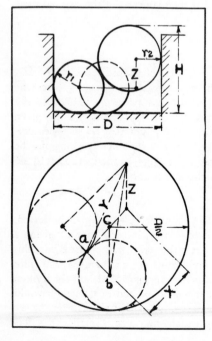

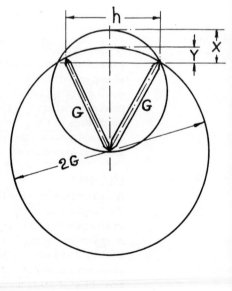

Figure 12.7.

A point to note is that, if the angle of the plane passing through the centres of the three balls is relatively low, then the accuracy of the bore measurement will be correspondingly high. Suppose that this plane makes an angle of $30°$ to the horizontal; an inaccuracy of, say, 0.06 mm in the height measurement would result in an inaccuracy in the bore size of only 0.03 mm. By measuring the height dimension using a comparator with a flat tip, the method can be very precise.

EXAMPLE 7: MEASUREMENT OF LARGE BORES

Inspection can be carried out by the use of bar gauges with rounded ends, the amount of side play found being an indication of the extent to which its diameter is too large. In Figure 12.7 if the side movement h is estimated using a rule, then the error can be determined by the use of the theorem of intersecting chords, where G represents the length of the gauge. The error is $X - Y$ and in the larger circle $(2G - Y)Y = (h/2)^2$ and as Y will be small

$$2GY \approx \frac{h^2}{4}$$

$$Y = \frac{h^2}{8G}$$

In the smaller circle

$$(G - Y)X = (\frac{h}{2})^2$$

and

$$GX \approx \frac{h^2}{4}$$

thus

$$X = \frac{h^2}{4G}$$

and the error is $h^2/4G - h^2/8G$. Therefore the error is $h^2/8G$, where h is the amount of side play and G is the gauge size. One advantage of this method is that a direct estimate is made of a small quantity, the amount of error. This means that any given percentage of error in estimating the side play by rule measurement leads to the same percentage error when working out the amount by which the bore is too large; so the method is by no means as inaccurate as might at first appear.

EXAMPLE 8

An alternative method is given in Figure 12.8(a). As indicated, the length L is a very small amount and less than D, giving a small amount of rocking on either side of the centre line; this is indicated by w. The conditions are exagerrated in Figure 12.8(b). The full circle is the hole being gauged, and the broken circle with centre A is that which the end of the gauge would describe if it made a full sweep. Actually it moves over the arc BHF, and the amount by which the gauge is smaller than D is

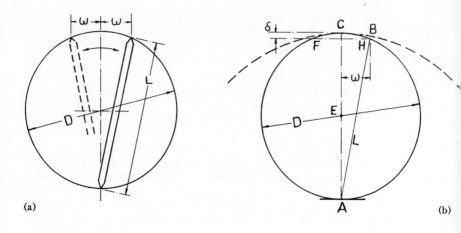

(a)
(b)

Figure 12.8.

shown by CH. Let this be δ.

A solution accurate enough for most practical purposes is:

If BC is joined, angle B is a right angle, since it is the angle in a semi-circle, and, in triangle ABC, $AC^2 = CB^2 + AB^2$. $AC = D$, $AB = L$ and CB is very nearly equal to w. Hence

$$D^2 \approx L^2 + w^2$$

But
$$D = L + \delta$$
$$(L + \delta)^2 = L^2 + w^2$$
$$L^2 + 2L\delta + \delta^2 = L^2 + w^2$$

δ^2 will be so small that it can be ignored. Hence

$$L^2 + 2L\delta = L^2 + w^2$$
$$2L\delta = w^2$$
$$\delta = \frac{w^2}{2L}$$

EXAMPLE 9: PROBLEM OF MEASURING A RELIEVED BORE

The bore (Figure 12.9(a)) does not have two points on a common diameter. The workpiece is mounted on a measuring machine. A ball feeler of diameter $2r$ contacts the bore at M and N successively, and the distance AB through which the feeler is displaced is recorded. Next, the feeler is displaced through a known distance E perpendicular to AB, and the length CD is measured, the ball making contact at points T and S.

Lengths AB, CD and E are known and also the radii OM and OT; also the perpendicular FH can be drawn. Then $AF = AB/2$ and $CH = CD/2$. The triangles OFA and OHC are right angled; hence

$$OA^2 = AF^2 + OF^2 \qquad (12.1)$$
$$OC^2 = CH^2 + OH^2 \qquad (12.2)$$

Also $\qquad OA^2 = OC = R - r$

Since OA = OC, from the equations (12.1) and (12.2)

$$AF^2 + OF^2 = CH^2 + OH^2$$

However, $\qquad OH = E + OF$

and, by substitution,

$$AF^2 + OF^2 = CH^2 + E^2 + OF^2 + 2E \times OF$$

and it follows that

$$OF = (AF^2 - CH^2 - E^2)/2E$$

Now, CH = CD/2 and AF = AB/2, so that

$$OF = (AB^2 - CD^2)/8E - E/2$$
$$= (AB + CD)(AB - CD)/8E - E/2$$

From equation (12.1),

$$OA = (AB^2/4)^{1/2} + OF^2$$

and from the figure it can be seen that

$$R = OA + r$$

Precise measurement of AB and CD, separated by an arbitrary distance E, can be carried out more readily on a measuring machine, but it is more difficult when using slip gauges and rollers. In this case measurement can be simplified by using the inverse of the technique described. Instead of measuring AB and CD, arbitrary values can be assigned to these lengths. The distance E is then measured by feeler, at centre distances of AB and CD, make contact with the bore at points M and N, and T and S.

For this operation a pair of gauges (Figure 12.9(b)) can be used. It is of assistance if the distance AB is a whole number of millimetres with the diameter $2r$ of the rollers known accurately and the distance a identical for both gauges. The

Figure 12.9.

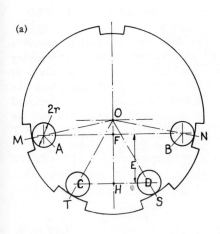

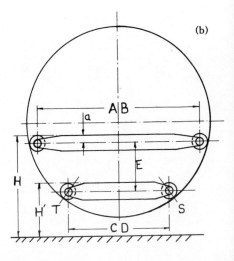

component is placed on a surface plate with the axis of the bore horizontal. Height H is then measured, the gauge AB removed and the process repeated using the second gauge with a centre distance of CD to obtain a value for the distance H'.

Since dimension a is common, the value $E = H - H'$., and a value for OF can be calculated from the equation

$$OF = (AB + CD)(AB - CD) / 8E - E/2$$

and the radius of the bore obtained from

$$R = (AB^2/4 + OF^2)^{1/2}$$

The minimum diameter that can be measured is greater than AB + 2r.

EXAMPLE 10: CHECKING PERIPHERY (BRAKE DRUM)

The radius of arcs of circles can be measured by the method shown in Figure 12.10. Parallel strips and rollers are used, the centre distance being known. The component rests on two rollers with a small gap between the circumference of the arc and surface plate. The gap can be measured with slip gauges, and the radius can be evaluated. From Pythagoras' theorem

$$AB^2 = BC^2 + AC^2$$

and

$$AB = R + r$$
$$BC = R + b - r$$
$$AC = L/2$$

Therefore

$$(R + r)^2 = (R + b - r)^2 + (L/2)^2$$
$$R^2 + 2Rr + r^2 = R^2 + 2bR - 2Rr - 2br + b^2 + r^2 + L^2/4$$

Thus

$$4Rr - 2Rb = L^2/4 - 2rb + b^2$$

and

$$2R(2r - b) = L^2/4 - 2rb + b^2$$

Therefore

$$R = \frac{L^2/4 - 2rb + b^2}{2(2r - b)}$$

and

$$R = \frac{L^2}{8(2r - b)} - \frac{2rb - b^2}{4r - 2b}$$

Figure 12.10.

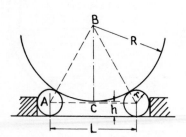

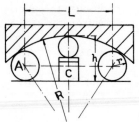

Figure 12.11.

Therefore

$$R = \frac{L^2}{8(2r-b)} - \frac{b}{2}$$

EXAMPLE 11: CHECKING INTERNAL RADIUS

The same method of calculation can be used, the formulae varying little from that of the previous example. The set-up is shown in Figure 12.11; the rollers and slips are used to determine the distance from the arc to the surface plate. The abbreviated calculations are as follows:

$$AB = R - r$$
$$BC = R - b + r$$

and $\quad AC = L/2$

Thus

$$(R - r)^2 = (R - b + r)^2 + L^2/4$$

Therefore

$$2Rb - 4Rr = L^2/4 - 2br + b^2$$

$$R = \frac{L^2}{8(b - 2r)} + \frac{b}{2}$$

EXAMPLE 12: CHECKING ARCUATE SURFACE OF LARGE RADIUS

This is an application of the sine principle. The workpiece X (Figure 12.12(a)) is set up on a sine table Y, and, to check that the radius is correct, points A, B and C are chosen so that each point lies in a plane that is perpendicular to the pivot axis H.

Initially the angles r, s and t are calculated; these correspond to the points A, B and C, and the radius OA is assumed correct. The work is positioned so that the centre O is on a perpendicular to the base RS which passes through the pivot H.

The distance a is measured, i.e. the distance from the base RS to a tangent to the arcuate surface that is parallel to the base. The dimension b (equal to $a - d$) is then calculated. The platen of the sine table is next tilted to the angle r by supporting the free end on a stack of slip gauges of the required height (Figure 12.12(b)). Then the workpiece is slid on the platen until a tangent to the arcuate surface, passing through the point A and parallel to the base RS, is at a distance m from the base. The distance

$$m = d + AD$$

and since $\quad AD = OA$

$$OD = OH \cos r$$

and $\quad OH = OA - b$

then

$$m = d + OA(1 - \cos r) + b \cos r$$

Next, the platen of the sine table is inclined successively to the angles s and t of Figure 12.12(c) and Figure 12.12(d). The distances n and p are measured, i.e. from the base to tangents passing through the points B and C. Then, $n - d = $ BE and $p - d = $ CF. If the radius of the arcuate curve is correct,

$$BE = OA(1 - \cos s) + b \cos s$$

and

$$CF = OA(1 - \cos t) + b \cos t$$

Figure 12.12.

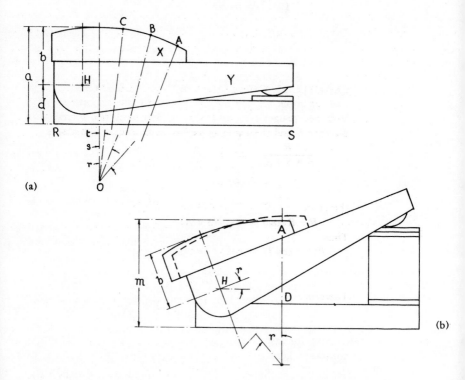

(a)

(b)

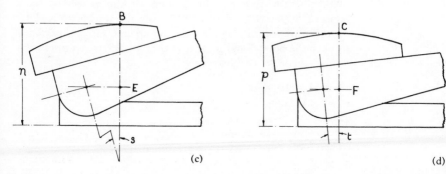

(c)

(d)

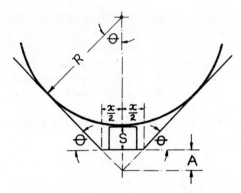

Figure 12.13.

EXAMPLE 13: CHECKING ANGULAR SURFACES
This example (Figure 12.13) refers to the calibration of a
V-shaped gauge; the problem is to determine the intersection
distance X and the angle, as well as to determine the radius.

$$\frac{R}{R + S + A} = \cos\theta$$

$$A = \frac{X}{2}\tan\theta$$

therefore $\quad R = R\cos\theta + \cos\theta\,(S + A)$
$\quad\quad R - R\cos\theta = \cos\theta\,(S + A)$

Thus

$$R(1 - \cos\theta) = \cos\theta\,(S + A)$$

$$R = \frac{\cos\theta\,(S + A)}{1 - \cos\theta}$$

Therefore

$$R = \frac{1}{\sec\theta - 1}\,(S + A)$$

Now for ease of use $1/(\sec\theta - 1)$ and A can be made simple
constants. Let us assume, for instance, a formula

$$R = 10(S + 10)$$

was considered desirable; then

$$1/(\sec\theta - 1) = 10 \quad \text{Therefore}$$
$$10(\sec\theta - 1) = 1$$
$$\sec\theta - 1 = 0.1$$
$$\sec\theta = 1.1$$
$$\theta = 24° 38'$$

Now $A = 10$, then

$$(X/2)\tan\theta = 10$$

thus

$$X = 20\cot\theta$$

Therefore

$$X = 20\cot 24° 38' = 2.1808 \times 20 = 43.616\text{mm}$$

Alternatively, let us assume a formula
$$R = 5(S + 5)$$
Then
$$1/(\sec\theta - 1) = S$$
$$\sec\theta - 1 = 0.2$$
$$\sec\theta = 1.2$$
$$\theta = 33°33'$$
Now
$$(X/2)\tan\theta = 5$$
Therefore
$$X = 10\cot\theta = 10\cot 33°33' = 15.137 \text{ mm}$$

EXAMPLE 14: MEASUREMENT OF INTERSECTION DISTANCE X

The angles are unequal on the V of Figure 12.14. The angles θ and ϕ have been previously measured. Measurement is by using slips and dial test indicator. The roller diameter is 25 mm; $\theta = 45°50'$; $\phi = 29°10'$; $H = 110.89$ mm; $\Delta = (\theta + \phi)/2 = 37°30'$. Now
$$A = r\operatorname{cosec}\Delta = 12.5\operatorname{cosec}37°30'$$
Thus
$$A = 20.537 \text{ mm}$$
$$B = A\cos\alpha \text{ and } \alpha = \theta - \Delta = 8°20'$$
Therefore
$$B = 20.537\cos 8°20' = 20.319 \text{ mm}$$
Now
$$X = H - (r + B) = 110.89 - (12.5 + 20.319)$$
$$= 78.071 \text{ mm}$$

EXAMPLE 15: LOCATION OF POSITION OF INTERSECTION A

The gauge is shown in Figure 12.15, and it is required to

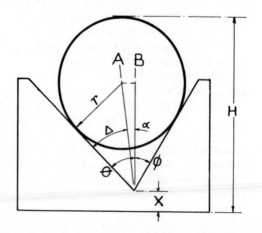

Figure 12.14.

calculate the value of the check dimension M. $\theta = 35°$; the roller diameter is 28 mm; the slips are 24 mm.

$$M = 14 + X + Y + 22 = 36 + X + Y$$
$$X = 14 \sec 35° = 17.091$$
$$Y = 17 \tan 35° = 11.904$$

Therefore

$$M = 36 + 17.091 + 11.904 = 64.995 \text{ mm}$$

Figure 12.15.

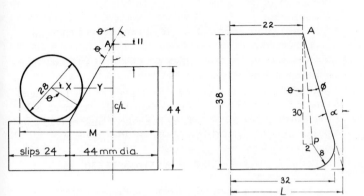

Figure 12.16.

EXAMPLE 16

In this example we determine the angle α and the intersecting distance L on the gauge shown in Figure 12.16.

$$\alpha = \theta + \phi$$
$$\tan \theta = 2/30 = 1/15$$

Therefore

$$\theta = 3°49'$$
$$\sin \phi = 8/2 \operatorname{cosec} 3°49' = 4\sin 3°49'$$
$$= 4 \times 0.0666 = 0.2663$$

therefore

$$\phi = 15°26'$$
$$\alpha = 3°49' + 15°26' = 19°15'$$

To find L, we use

$$(L - 22)/38 = \tan \alpha$$

and therefore

$$L = 38 \tan \alpha + 22 = (38 \times 0.3297) + 22$$
$$= 34.528 \text{ mm}$$

EXAMPLE 17: CHECKING SQUARES

Indicate by diagrams and descriptions three methods of checking the squareness of components showing the importance of the right angle representing one-quarter of a revolution.

The squareness of the ends of a block and the parallelism between its faces can be directly measured by the NPL indicator shown in Figure 12.17(a). The stop near the base is pressed up against the surface to be inspected, and an indicator reading is made. Readings are made for all the four angles of the block. The average of these angles must represent a right angle. Thus the mean of the four readings represents true squareness whatever the individual angles may be. The tester can be presented to a nominal 90° angle, and the out-of-squareness can be determined by the difference between the reading obtained and the true squareness reading.

The accuracy of a square (Figure 12.17(b)) can be measured by two discs of the same diameter attached to an angle plate. They should be set so that two equal slip blocks are held in position between the blade of the square and the discs. The square is then moved to the other side of the discs when, if the square is accurate, the two slip blocks will be held in position again. Should the square be inaccurate, the size of one of the slips must be changed until the slips are held as before. The difference in size between the two slips represents twice the error.

An alternative method is to use an auto-collimator set up horizontally on a surface plate with a slip block held against the vertical face to be tested, readings being taken for the four angles.

Figure 12.7(c) shows the use of an optical square which is so designed that rays of light are always deviated through a right angle. The set shown is for checking the squareness of a milling machine table and a column. A reading on the auto-collimator is obtained from the reflector A on the table. The optical square is set against the face of the column, and the reflector is positioned on the vertical face of the column. If the work table and the column are square, then no change of reading will be observed through the telescope in the two positions. For working on vertical faces the magnetic type of base support is useful.

Figure 12.17.

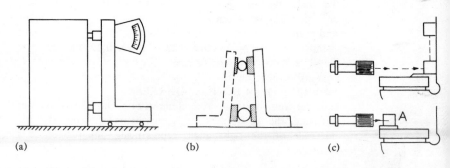

(a)   (b)   (c)

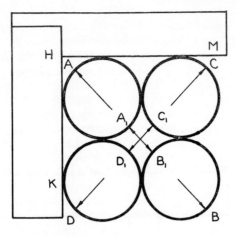

Figure 12.18.

EXAMPLE 18: TESTING A SQUARE BY MEANS OF DISCS
Describe the means of checking for accuracy of a right angle using four discs.

Figure 12.18 shows the set-up with the discs in position. Test the distances AB and CD or AB and CD. If the two distances measure the same, the line KH and HM must form a right angle. Any deviation will be multiplied by 2, making this test very accurate. The test does not prove anything except the position of the points touched by the discs, but in practical square making the first thing is to get the blade and stock as near to straight edges as possible.

Button and Disc Location

It is required to bore the two holes in the body of a gear pump using tool makers' buttons for setting up the casting on a lathe faceplate. Describe the procedure, the bores being 100 mm centres.

EXAMPLE 19
Mark out the centres, and drill and tap two holes for the button screws at about the correct centres; clamp the two buttons 12 mm in diameter loosely on the casting. Note that holes in buttons are about 4 mm larger than screws. A micrometer is set to measure 112 mm over the two buttons which are lightly tapped to give this dimension and then are clamped by the screws. The casting is now bolted to the lathe faceplate with one button as near as possible to the centre, until by revolving against a dial indicator it registers true. The button is now removed, and the tapped hole is drilled away. It is important that the hole is enlarged by a single-point tool and not by a drill which would follow the tapped hole. Boring then proceeds until size is obtained, and the second button is set in the same way. The set-up is shown in Figure 12.19.

Figure 12.19.

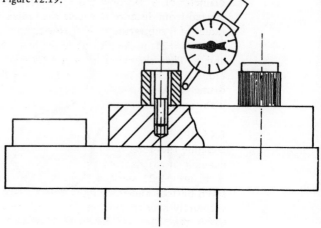

EXAMPLE 20: LOCATION OF WORK BY DISC METHOD

Figure 12.20(a), (b) shows an example of precision work. Explain how discs can be used instead of buttons to obtain exact centre distances to the dimensions given.

Three discs must be made to such diameters that, when their peripheries are in contact, each centre will coincide with the hole centre to be bored. The diameters can be found as follows. Subtract dimension Y from X, thus obtaining the difference between the radii of discs C and A (Figure 12.20(b)). Add this difference to dimension Z, and the result will be the

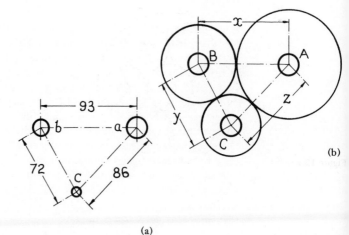

(b)

Figure 12.20.   (a)

diameter of A. Dividing this by 2 gives the radius, which sub-tracted from distance X equals the radius of B. Similarly, the radius of B subtracted from Y equals radius of C. For example, 93 − 72 = 21 or the difference between the radii of discs C and A. Then the diameter of A is 21 + 86 = 107 mm and the radius 53.5 mm. The radius of B is 93 − 53.5 = 39.5 and the diameter 79 mm. The centre distance 72 − 39.5 = 32.5, which is the radius of C and so the diameter of C is 65 mm.

EXAMPLE 21
Describe the 'three-disc method' of locating holes, and give an example of circular spacing.

Figure 12.21 shows a method of locating six holes using seven buttons and three discs. The bores of the discs must fit accurately over the buttons. First the centre button is screwed to the template, and the disc A is slipped over it. Next disc B is placed in contact with A, and its button is inserted. The third disc C next contacts A and B by which the third button can be set in position, and so on until all seven buttons are in position and the template is ready for boring.

Suppose it is required to locate nine equally spaced holes on a circumference 180 mm in diameter. The size of the smaller disc (Figure 12.21) is found by multiplying the diameter of the circle upon which the centres of the discs are located by the sine of half the angle between the two adjacent discs. This angle equals 360° divided by the number of discs, i.e. 360/9 = 40; hence the diameter of the smaller disc is $180 \times \sin 20° = 180 \times 0.34 = 61.20$ mm. Then 180 − 61.2 = 118.8 which is the diameter of the central disc.

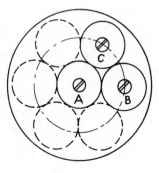

Figure 12.21.

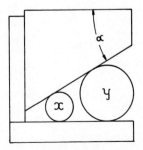

Figure 12.22.

EXAMPLE 22: ACCURATE ANGULAR MEASUREMENT WITH DISCS

If the required angle is $32°$ the radius of the large disc is 50 mm and that of the small disc is 25 mm, what is the centre-to-centre distance?

From Figure 12.22, the sine of half the required angle is 0.275. The difference between the radii of the discs is 50 − 25 = 25, and 25/0.275 = 90.9 mm, which is the centre distance.

Given now that the required angle α is $15°$ and that the discs are required to be in contact, find the diameter of the large disc if the small one is 50 mm in diameter.

The sine of $7° \, 30'$ is 0.130. Multiplying twice the diameter of the small disc by this value, we obtain 2 × 50 × 0.130 = 13 mm. This product divided by 1 minus the sine of $7° \, 30'$ is 13/(1 − 0.130) = 14.90. This quotient added to the diameter of the small disc is 14.90 + 50 = 64.90 mm, which is the diameter of the large disc.

Figure 12.23.

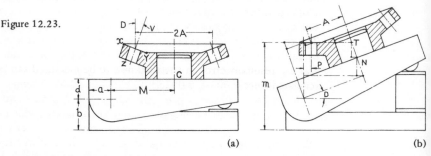

(a)   (b)

Problems of the Sine Principle

EXAMPLE 23

Figure 12.23(a) shows a component with two holes bored at right angles to the circular conical surface XY. It is required to determine the distance 2*A* between the points of intersection of the axes. As shown, spigot C is mounted on the sine table, so that its axis is at a known distance *M* from the pivot position. The distances *b* and *d* are known.

As indicated in Figure 12.23(b) the sine table is inclined so that the surface XY and the axis VZ of the hole are parallel and perpendicular to the base of the table. It is necessary to know the distance TN. The height *m* from the conical surface to the base of the table is measured, then

$$TN = (m - b)/\cos D - (d + M\tan D)$$

Next, to find the dimension *P*, measure the distance from the centre of the hole to the end datum face of the sine table, and subtract the known distance *a*. It can be seen that

$$A = BT \cos D$$

Since

BT $= M\cos D - P - (\text{TN} + d)\sin D$ (12.3)

then

$A = M\cos^2 D - P\cos D - d\sin D\cos D - \text{TN}\sin D\cos D$ (12.4)

If TN in equation (12.4) is replaced by its value from equation (12.3),

$$A = M\cos^2 D - P\cos D - d\sin D\cos D - mD + b\sin D + d\sin D\cos D + M\sin^2 D$$

$$= M(\sin D + \cos^2 D) - P\cos D - m\sin D + b\sin D$$

Since $\sin^2 D + \cos^2 D = 1$

$$A = M - P\cos D - (m - b\sin D)$$

EXAMPLE 24: USE OF SINE TABLE FOR LINEAR MEASUREMENT

From the set-up in Figure 12.24 it is required to measure the thickness z at the zero mark on the tapered block A, the taper being 0.0025 in./in. or $0° 3' 26''$.

The block is mounted on a sine table; the dimensions b, c and d are known. It is necessary to set the zero mark of the block A accurately in line with the axis XY of the pivot of the sine table, and this setting can be effected with slip gauges. The combined length of these is $h = b - g$. The dimension g can be readily measured accurately to within 0.004 in. (0.1 mm), and the resultant error in the thickness to be measured can be regulated since 0.004×0.0025 is 0.00001 in. (0.0002 mm).

The table is tilted to the angle M, i.e. $\tan^{-1} 0.0025$ equal to $3' 26''$ by means of slip gauges V, the height of which is $b\sin M$. The face PN is then parallel to the base RS of the table, and the height m can be measured without difficulty. The thickness z of the tapered block is then equal to $(m - c)/\cos M - d$.

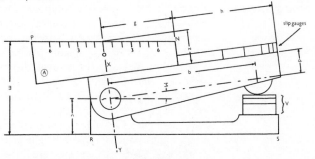

Figure 12.24.

EXAMPLE 25: CHECKING DIMENSIONS OF SPLINED SHAFTS

Figure 12.25 illustrates a splined shaft. It is necessary to know the correct theoretical overall dimension M, and it is required to derive the formula.

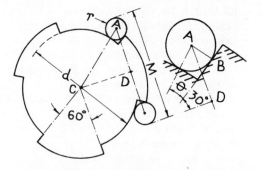

Figure 12.25.

There are three equally spaced splines and the accuracy of the $60°$ angles may be checked using two rollers and a micrometer. It is first necessary to check the actual value of the diameter d, and the shaft should be placed in bench centres, a setting being obtained with an indicator reading over a bar of known diameter concentric with its centres and whose size is near to the nominal diameter d of the component. The radius of this diameter can be measured according to the indicator reading obtained when compared with the reading recorded on the test bar. The shaft is then checked by making a micrometer measurement over two suitable rollers as shown. The overall dimension M is given as

$$M = 2(AD + r)$$

Now

$$\sin \theta = \frac{AB}{AC} = \frac{r}{d/2 + r}$$

from which θ may be found. Also

$$\sin(\theta + 30) = AD/AC$$

Therefore

$$AD = AC \sin(\theta + 30) = (d/2 + r) \sin(\theta + 30)$$

Thus

$$M = 2[(d/2 + r) \sin(\theta + 30) + r]$$

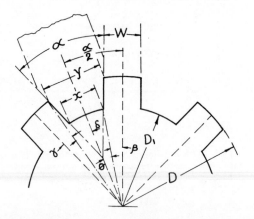

Figure 12.26.

EXAMPLE 26: SPLINE CUTTER

Figure 12.26 shows the method of laying out a spline cutter. Calculate the required dimensions for making the cutter.

$$\sin \beta = W/D_1$$
$$\theta = a/2 - \beta$$
$$X = D_1 \sin\theta$$

which is the dimension across the sharp corners of the cutter. Similarly to find the checking dimension Y (which may be measured with gear tooth calipers)

$$\sin \gamma = W/D$$
$$\delta = a/2 - \gamma$$

therefore

$$Y = D \sin\delta$$

which is the dimension at the top of the sphere.

EXAMPLE 27: MEASUREMENT OF SPLINE SHAFT CUTTER

The measurement involves verification within given tolerances of the arc D, and checking the values of angles α (Figure 12.27 (a), (b)). In practice it is convenient to measure the dimension M which equals $2AF$, where

$$AF = AC/\cos\alpha$$

and

$$AC = AE - CE = (D \sin\alpha - B)/2$$

so

$$M = (D \sin\alpha - B)/\cos\alpha = D \tan\alpha - B \sec\alpha$$

If the cutter is used to produce splines of width B within limits B_2 and B_1, the magnitude of dimension M must lie between the maximum and minimum values of M_1 and M_2, corresponding to the extreme values D_2 and D_1 allowed for the diameter D. The minimum value M_2 will occur when B and D are maxima since, with the cutter at maximum diameter D_1, the width of the splines produced must be equal to, or smaller than, the upper limit B_1, and the value of M_2 is then given by

$$M_2 = (D_1 \sin\alpha - B_1)/\cos\alpha$$

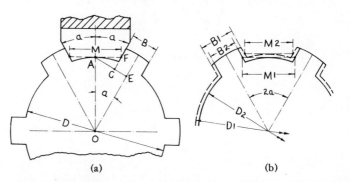

Figure 12.27.

(a)   (b)

Similarly, the maximum value M_1 can be obtained from the expression

$$M_1 = (D_2 \sin \alpha - B_2)/\cos \alpha$$

Determine the limiting dimensions M_1 and M_2 for a cutter to mill six splines of width 10 mm to a spline root diameter of 16 mm, the tolerances being -0.016 and -0.034 mm; so the maximum and minimum values of D are 15.984 mm (D_1) and 15.966 mm (D_2). For a width of 10 mm the limits are -0.030 and -0.078 mm, giving

$$B_1 = 3.970 \text{ mm}, \quad B_2 = 3.922 \text{ mm}$$

For a shaft with six splines, angle α is $30°$, so that

$$\cdot M_2 = (15.984 \sin 30 - 3.970)/\cos 30$$
$$= 4.644 \text{ mm}$$

and

$$M_1 = (15.966 \sin 30 - 3.922)/\cos 30$$
$$= 4.689 \text{ mm}$$

Inspection Using Co-ordinate Dimensions

EXAMPLE 28

The locating, boring or checking of holes on either jig-boring machines or inspection machines requires that the operator establishes 'zero points' from some basic points on the workpiece. Indicate the difference in the dimensioning system using co-ordinate figures from one using standard dimensions.

Consider Figure 12.28. The distances indicated by X would be standard toleranced sizes as given by the designer while the distances Y are established from the two zero points. These are basic values and do not show working tolerances, but they

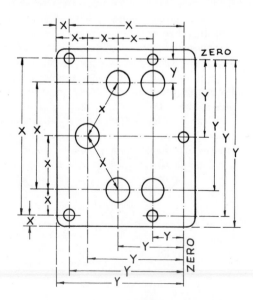

Figure 12.28.

are the dimensions the operator uses to set the measuring instruments. Having located one hole, the positions of other holes relative to the first must be determined by reference to two centre lines at right angles. The distances of a point from two axes at right angles are called its co-ordinates, and from this definition the calculation derives its name.

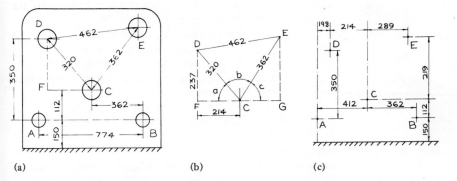

(a)   (b)   (c)

Figure 12.29.

EXAMPLE 29

Calculate suitable co-ordinate dimensions for boring or inspecting the various holes in the gear box of Figure 12.29(a). There is a machined face at the bottom of the box, and the holes are indicated by A, B, C, D and E.

The co-ordinates for A, B and C are given, so only D and E need be considered. Considering D, the vertical distance DF from C to a horizontal centre line through D is given by

$$DF = 350 - 112 = 238 \text{ mm}$$

This together with the centre distance of 320 mm gives the perpendicular and hypotenuse of triangle CFD:

$$CF^2 = \quad CF^2 = 320^2 - 238^2 = 460 \text{ mm}$$
$$CF = 460^{1/2} = 214 \text{ mm}$$

From this the horizontal distance from A to F will be 775 - 365 - 214 = 198 mm.

E can be found from the detail diagram in Figure 12.29(b), where EG and CG are vertical and horizontal lines, and their lengths are required. If angle c is found, then with CE given it will be possible to find EG and CG. In triangle DFC,

$$\frac{DF}{FC} = \tan a = \frac{237}{214} = 110$$
$$a = 47° 56'$$

Applying the cosine rule to triangle CDE, then three sides being given

$$DE^2 = DC^2 + CE^2 - 2 \times DC \times CE \cos b$$
$$462^2 = 320^2 + 362^2 - 2 \times 320 \times 362 \cos b$$
$$214 = 102 + 131 - 232 \cos b$$

$$\cos b = \frac{102 + 131 - 214}{232} = 0.085$$
$$b = 85°5'$$

Since

$$a + b + c = 180°$$
$$c = 180 - (a + b) = 180 - (47°56' + 85°5')$$
$$= 46°54'$$

Now

$$EG = EC\sin c = 362\sin 46°59' = 362 \times 0.60$$
$$= 218\,\text{mm}$$

Also

$$GC = 362\cos c = 362 \times 0.797 = 289\,\text{mm}$$

The second detail diagram (Figure 12.29(c)) shows where the co-ordinate dimensions would be located.

EXAMPLE 30: DIMENSIONING A JIG PLATE

Indicate a means by which the plate can be produced if the dimensions are produced using polar co-ordinates. The operation is that of boring the six holes (Figure 12.30).

Using a co-ordinate method as described, it would be necessary to calculate the dimensions A, B, C, D, E, F and G. If the machine is fitted with a circular table, the holes may be bored from the centre of the circle, and the only dimension required will be the diameter of the pitch circle H of the holes and the angular spacing between them, i.e. 60°. It is recommended that a central hole J is provided for setting purposes; this hole is known as a reference hole and simplifies checking of the jig after manufacture. For checking, the chord K should be given as this proves useful whatever method of boring is employed.

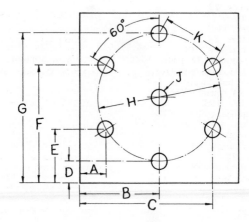

Figure 12.30.

INDEX